Lecture Notes in Mathematics

Edited by A. Dold and B. Eckmann

610

Gary R. Jensen

Higher Order Contact of Submanifolds of Homogeneous Spaces

Springer-Verlag
Berlin Heidelberg New York 1977

Author
Gary R. Jensen
Department of Mathematics
Washington University
St. Louis, Missouri 63130
USA

AMS Subject Classifications (1970): 53A05, 53A15, 53A20, 53A40, 53A55

ISBN 3-540-08433-9 Springer-Verlag Berlin Heidelberg New York
ISBN 0-387-08433-9 Springer-Verlag New York Heidelberg Berlin

Printing and binding: Beltz Offsetdruck, Hemsbach/Bergstr.
2141/3140-543210

TABLE OF CONTENTS

Page

Introduction V

I. The general theory 1

1. Contact 1
2. The Grassmann bundle 5
3. Adapted frames of a homogeneous space 6
4. Zeroth order frames of a submanifold 7
5. Lie transformation groups 8
6. First order frames 9
7. Second order frames 12
8. Third and higher order frames 17
9. Frenet frames 22
10. The role of the Maurer-Cartan form of G 23
11. Congruence and Existence Theorems 30
12. Homogeneity Theorem 41

II. Surfaces in $\mathbb{R}^3$ under $\mathbb{E}(3)$ 44

III. Real curves in $\widetilde{G}_{4,2}$ under $SO(4)$ 58

IV. Holomorphic curves in $\mathbb{C}P^2$ under $SU(3)$ 66

V. Holomorphic curves in $\mathbb{C}G_{4,2}$ under $SU(4)$ 81

VI. Surfaces in $\mathbb{R}^3$ under the special affine group 115

References 151

Index 152

INTRODUCTION

These notes contain an exposition of Elie Cartan's theory of higher order invariants of submanifolds of homogeneous spaces treated by the method of moving frames. The theory is then applied to five examples. We were introduced to Cartan's magnificent book [C, 1937], and our view of the theory is strongly guided, by the paper of P. Griffiths [G, 1974].

It is Cartan's stated aim in his book (which consists of the lecture notes taken by Jean Leray of a course given by Cartan at the Sorbonne in 1931-32!) to develop the fundamental theorems of Lie groups together with the basic principles of the method of moving frames in differential geometry. They are developed together because the latter depends in such an essential way on the former. For example, the classical equations of Darboux in the theory of surfaces in Euclidean space are just the structure equations of the group of Euclidean motions. In more general Klein-geometries, where the Euclidean motions are replaced by an arbitrary Lie group G, Darboux's structure equations are replaced by the structure equations of G, which were first established by Maurer in 1888.

The basic theory of Lie groups - the Maurer-Cartan forms and equations, the theorems of Lie, etc. - is now standard knowledge in differential geometry, and in many other fields as well; but the method of moving frames is still not widely known (except perhaps in name), with the very notable exception of S.S. Chern's repeated use of the method for decades. For this reason these notes deal with the method of moving frames as applied to submanifolds of homogeneous spaces, and assume the basic material of Lie group theory.

In his review [W, 1938] of Cartan's book, Hermann Weyl observes that one may consider two different problems: (1) the fixed parametrization problem in which the submanifold is given with a fixed parametrization; and (2) the unparametrized problem. In modern terminology these problems may be described as follows. Answer questions of contact, congruence, existence and homogeneity for: (1) imbeddings $f, \tilde{f}: S \to M$, where S is a given p-manifold; and (2) imbeddings $f: S \to M$ and $\tilde{f}: \tilde{S} \to M$, where $\tilde{S}$ and S are given p-manifolds.

In his paper [G, 1974], Griffiths considers problem (1), but following Cartan we have dealt with problem (2) in these notes. The problems are not altogether independent, of course, as we shall try to elucidate below, but first back to Weyl (p. 601 of [W, 1938]).

> It is this problem with which Cartan deals in the present book, and in some way he reduces the second influence, the choice of parameters, to the choice of the frame. I do not quite understand how he does this in general, though in the examples he gives the procedure is clear. To me it seems advisable to keep both factors apart from the beginning; the process itself tends to normalize both in mutual interdependence as it advances to higher and

> and higher orders. The same situation is met with everywhere in differential geometry. For instance, riemannian spaces could be treated by introducing coördinates and attaching to each point $\underline{A}$ a frame, that is, a cartesian set of axes. Invariance is required with respect to arbitrary transformations of the coördinates and to orthogonal transformations of the frames which may depend arbitrarily on the point $\underline{A}$. One knows how Gauss, Riemann, and Einstein got around the frames: the parameters once chosen define uniquely at each point an affine set of axes, and one takes advantage of it by treating cartesian geometry in terms of affine frames and a fundamental metric form rather than in terms of cartesian frames. Cartan goes here to the opposite extreme by normalizing the parameters in terms of frames. I should advocate full impartiality on both sides as long as one deals with fairly general differential geometric problems.*

The difference between problems (1) and (2) can be easily seen for the notions of first order contact and congruence. If $\tilde{f}, f: S \to M$ are imbeddings, then in the fixed parameter sense they have at least first order G-contact at s if $f_*(s) = A_* \circ \tilde{f}_*(s)$ as linear maps $S_s \to M_{f(s)}$, for some $A \in G$; and they are congruent if $f = A \circ f$ for some $A \in G$. On the other hand, imbeddings $f: S \to M$ and $\tilde{f}: \tilde{S} \to M$ have at least first order G-contact at $s \in S$ and $\tilde{s} \in \tilde{S}$ in the unparametrized sense, if $f_*(S_s) = A_* \tilde{f}_*(\tilde{S}_{\tilde{s}})$ for some $A \in G$, while they are congruent in this sense if $f(S) = A\tilde{f}(\tilde{S})$ for some $A \in G$

The main idea of the method of moving frames can be roughly described for the first order case as follows. In problem (1) it consists of choosing a "normal form" for the linear map $f_*: S_s \to M_{f(s)}$

---

with respect to all possible "adapted" frames of M at f(s) (and the fixed parameter frame of S_s), where "adapted" frames of M are those frames of M coming from translation by elements of G of a fixed frame at a fixed origin. Thus the adapted frames at a point come from the linear isotropy action of G at that point applied to a single adapted frame there, and therefore one looks for a normal form of $f_*(s)$ under the linear isotropy action on $M_{f(s)}$.

In problem (2) it consists of choosing a "normal form" of the p-dimensional subspace f_*S_s of $M_{f(s)}$ in the Grassmann manifold of p-planes through the origin in $M_{f(s)}$, on which the isotropy subgroup acts in the obvious way. In other words, one chooses a cross section of the orbits of the linear isotropy action on this Grassmannian. From this description, the second problem appears simpler than the first, contrary to Weyl's feeling that the second problem is "slightly more complicated."

These notes are primarily an exposition of the method of moving frames: how they are constructed, how they relate to contact; the generalized Frenet frames and their role in proving theorems about congruence, existence and homogeneity. This is all contained in Cartan's book, yet Weyl writes in his review:

> The book under review pursues a three-fold purpose: it contains (1) an exposition of the general theory of finite continuous Lie groups in a terminology adapted to its differential geometric applications; (2) a general description of the method of repères mobiles; and (3) its application to a number of important examples.... While the topics (1) and (3) are given in

> full detail, the central problem (2) emphasized by this review is but briefly dealt with, at the beginning of Chapter 10 more from the standpoint of transformation groups, at the beginning of Chapter 12 more form the abstract standpoint. In both chapters there follow further applications: curves in the affine and the projective planes, and arbitrary surfaces in E_3. With the last example, the only one concerned with manifolds of more than one dimension, the integrability conditions turn up; although their rôle in the theory of Lie groups is amply discussed, their general formulation as an intrinsic part of the existence theorem in the theory of repèrage is omitted.
>
> All of the author's books, the present one not excepted, are highly stimulating, full of original viewpoints, and profuse in interesting geometric details. Cartan is undoubtedly the greatest living master of differential geometry.... Nevertheless, I must admit that I found the book, like most of Cartan's papers, hard reading....*

Hopefully, then, these notes can serve as an aid for reading Cartan's book itself. The wealth of ideas in his book is truly awe inspiring, and no one should study moving frames without consulting the original source.

A brief description of the content of these notes follows.

The first chapter begins with a discussion of the notion of contact between submanifolds of a given manifold. It is presented in such a way as to, hopefully, indicate the rationale of the frame construction and, especially, to make clear what k^{th} order frames and invariants have to do with k^{th} order contact.

We then show the involvement of the group G , (for a homogeneous space G/H) , in the frame construction. In particular, the role of the Maurer-Cartan form of G in the calculation of the invariants is explained. We stress the importance of the Maurer-Cartan form of G in the actual determination of the invariants, a fact which is not immediately clear because it is not even mentioned in the description of the frame construction.

The first chapter is concluded with statements and proofs of Cartan's main theorems on existence, congruence, and homogeneity of submanifolds of homogeneous spaces. The proofs of the latter two theorems are trivial for real analytic submanifolds, but quite difficult for the non-analytic case. The proof of the homogeneity theorem is a constructive one, as illustrated in many of the examples, and shows how to immediately obtain the Lie subalgebra of the subgroup of G which leaves the submanifold invariant.

The method of moving frames is developed in the first chapter under the most general assumptions, that is for homogeneous spaces G/H , where H is a closed subgroup of the Lie group G . However, the whole construction is greatly simplified when H is compact. For this reason, it is my recommendation that the first chapter first be read under the assumption that H is compact, which is the case for the first four examples. Then the H non-compact case can be tackled with the aid of the fifth example.

The first example considered is that of surfaces in $\mathbb{R}^3$ under the Euclidean group of proper motions. This example is treated at greater length in Cartan's book, and is included here because its great familiarity to most readers may make it a useful vehicle for making concrete the abstract exposition of the general theory.

The second example is that of real curves in the real Grassmann manifold of oriented 2-planes in $\mathbb{R}^4$, denoted $\tilde{G}_{4,2}$, under the action of $SO(4)$. We include this example primarily because of its simplicity.

The third and fourth examples are of holomorphic curves in the complex projective plane, $\mathbb{C}P^2$, under the action of $SU(3)$, and in the complex Grassmann manifold of complex 2-planes in $\mathbb{C}^4$, $\mathbb{C}G_{4,2}$, under the action of $SU(4)$. The second, third and fourth examples are contained in the paper [G, 1974] of Griffiths, but our treatments differ in many respects, sometimes substantially.

The fifth and final example is of special affine surface theory, that is, surfaces in $\mathbb{R}^3$ under the action of the special affine group. This is the only example considered for which the isotropy subgroup, here $SL(3;\mathbb{R})$, is non-compact, and consequently this is the only example in which some of the subtler aspects of the frame construction arise. In particular, the analysis of the Type IIb IIIb surfaces may prove to be the most useful example as an aid in understanding the general theory.

Each chapter has its own introduction. A classification scheme is introduced and explained at the beginning of Chapter II.

Throughout this paper the term smooth means "as differentiable as needed". For example, if contact of order $k \geq 1$ is being considered, then smooth maps are at least of class C^k . Hypotheses of real analyticity, i.e, class C^ω , will be explicitly stated.

The author wishes to express his gratitude to Washington University for granting him a Sabbatical leave for the current academic year, and to the University of California at Berkeley for the opportunity of spending this leave year with them. This work was accomplished in no small measure by the support received from these two institutions, which includes, of course, the help, encouragement and advice of many people in these universities. Special acknowledgement goes to Harold L. Raffill for the very high level of craftmanship he used in the typing of this manuscript.

I. The general theory.

I.1. Contact. Let M be a smooth m-dimensional manifold and let $f: S \to M$ and $\tilde{f}: \tilde{S} \to M$ be imbedded submanifolds of dimension p.

(0) f and $\tilde{f}$ have contact of at least order 0 at $s \in S$ and $\tilde{s} \in \tilde{S}$ if $f(s) = \tilde{f}(\tilde{s})$.

(1) f and $\tilde{f}$ have contact of at least order 1 at $s \in S$ and $\tilde{s} \in \tilde{S}$ if $f(s) = x = \tilde{f}(\tilde{s})$ and $f_* S_s = \tilde{f}_* \tilde{S}_{\tilde{s}}$ as subspaces of M_x.

We reformulate (1) in a way which naturally suggests how to define higher order contact. Let $G_{m,p}(M)$ denote the Grassmann bundle of tangent p-planes on M. Any immersion $h: N \to M$, where N is a p-manifold, induces a smooth mapping $T_h: N \to G_{m,p}(M)$ given by $T_h(s) = h_* N_s$, which is a p-plane in $M_{h(p)}$. Clearly, then, f and $\tilde{f}$ have contact of at least order 1 at $s \in S$ and $\tilde{s} \in \tilde{S}$ iff $T_f(s) = T_{\tilde{f}}(\tilde{s})$.

Observe that if $h: N \to M$ is an immersion, then $T_h: N \to G_{m,p}(M)$ is also an immersion, since $\pi \circ T_h = h$, where $\pi: G_{m,p}(M) \to M$ is the natural projection. Thus we may iterate this construction as follows: Let $M^{(0)} = M$, $m_0 = m$ and for $k \geq 0$ let $M^{(k+1)} = G_{m_k,p}(M^{(k)})$ and let $m_{k+1} = \dim M^{(k+1)}$. If $h: N \to M$ is an immersion of the p-manifold N into M, let $T_h^{(0)} = h$ and for $k \geq 0$ let $T_h^{(k+1)}$ $= T_{T_h^{(k)}}: N \to G_{m_k,p}(M^{(k)})$. Then for $k \geq 0$:

(k) f and $\tilde{f}$ <u>have contact of at least order</u> k at $s \in S$ and $\tilde{s} \in \tilde{S}$ if $T_f^{(k)}(s) = T_{\tilde{f}}^{(k)}(\tilde{s})$.

If f and $\tilde{f}$ have contact of at least order k at $s \in S$ and $\tilde{s} \in \tilde{S}$, but not of order $k+1$, then we say that f and $\tilde{f}$ have contact of order k at s and $\tilde{s}$.

<u>Example</u>

Suppose $p = 2$, S = a domain in $\mathbb{R}^2$, $M = \mathbb{R}^3$ and $f, \tilde{f}: S \to M$ are imbeddings given by $f(x, y) = (x, y, z(x, y))$, $\tilde{f}(x, y)$ $= (x, y, \tilde{z}(x, y))$, where z and $\tilde{z}$ are smooth functions on S. Then f and $\tilde{f}$ have contact at $s_0 = (x_0, y_0)$ of at least order 0 iff $z(s_0) = \tilde{z}(s_0)$, and they have contact of at least order 1 iff $z(s_0) = \tilde{z}(s_0)$ and $\operatorname{span}\{(1, 0, \frac{\partial z}{\partial x}(s_0)), (0, 1, \frac{\partial z}{\partial y}(s_0))\} = \operatorname{span}\{(1, 0, \frac{\partial \tilde{z}}{\partial x}(s_0$ $(1, 0, \frac{\partial \tilde{z}}{\partial y}(s_0)\}$. This latter equality holds iff $\frac{\partial z}{\partial x}(s_0) = \frac{\partial \tilde{z}}{\partial x}(s_0)$ and $\frac{\partial z}{\partial y}(s_0) = \frac{\partial \tilde{z}}{\partial y}(s_0)$.

As one would expect from a proper definition of kth order contact between the graphs of two functions, one can show that f and $\tilde{f}$ have contact at $s_0 \in S$ of at least order k if the kth order Taylor polynomials of z and $\tilde{z}$ are identical at s_0. We illustrate this for the case $k = 2$. Let $u = (\varepsilon_1, \varepsilon_2, \varepsilon_3)$ denote the global frame field on $\mathbb{R}^3$ given by the standard basis at each point. Using this global trivialization of $L(\mathbb{R}^3)$ we have $G_{3,2}(\mathbb{R}^3) \cong \mathbb{R}^3 \times G_{3,2}$, where $G_{3,2}$ is the Grassmann manifold of 2-planes in $\mathbb{R}^3$. Represent $G_{3,2}$ as $\mathbb{R}^{3\times 2^*}/GL(2;\mathbb{R})$, where $\mathbb{R}^{3\times 2^*}$ denotes the set of all 3×2 real matrices of rank 2 and $GL(2;\mathbb{R})$ acts on $\mathbb{R}^{3\times 2^*}$ by right matrix multiplication. Denote the equivalence class of $X \in \mathbb{R}^{3\times 2^*}$ by $[X]$. Now $f: S \to \mathbb{R}^3$ given by $f(x, y) = (x, y, z(x, y))$, induces a map $u_f: S \to G_{3,2}$ given by $u_f(s) = u^{-1} f_* S_s = \begin{bmatrix} 1 & 0 \\ 0 & 1 \\ z_x(0) & z_y(s) \end{bmatrix}$, and then $T_f: S \to G_{3,2}(\mathbb{R}^3)$ is given by $T_f(s) = (f(s), u_f(s)) \in \mathbb{R}^3 \times G_{3,2}$. The map $G_{3,2} \to \mathbb{R}^2$ given by $[{A \atop r\ t}] \to (r, t)A^{-1}$, defined on the open set $\{[{A \atop r\ t}] \in G_{3,2}: A \in GL(2;\mathbb{R})\}$, is a local coordinate map on $G_{3,2}$. Using it, together with the standard coordinates in $\mathbb{R}^3$, we get a local representation of T_f, namely $T_f = (x, y, z, z_x, z_y)$. Hence, with respect to these coordinates $T_{f*}S_{s_0}$

$$= \text{span}\{(1, 0, z_x(s_0), z_{xx}(s_0), z_{yx}(s_0)), (0, 1, z_y(s_0), z_{xy}(s_0), z_{yy}(s_0))\}$$

and it becomes clear that f and $\tilde{f}$ have contact at s_0 of at least order 2 iff the second degree Taylor polynomials of z and $\tilde{z}$ are identical at s_0.

The preceding example indicates how the following proposition is proved.

Proposition: Let $f\colon S \to M$ and $\tilde{f}\colon \tilde{S} \to M$ be smooth imbeddings, let $s_0 \in S$, $\tilde{s}_0 \in \tilde{S}$, and suppose $f(s_0) = 0 = \tilde{f}(\tilde{s}_0)$. Let $x^1, \ldots, x^m$ be a local coordinate system about 0 in M, so that $f = (f^1, \ldots, f^m)$ and $\tilde{f} = (\tilde{f}^1, \ldots, \tilde{f}^m)$, where $f^i = x^i \circ f$, $\tilde{f}^i = x^i \circ \tilde{f}$, $i = 1, \ldots, m$. Then f and $\tilde{f}$ have contact of at least order k at s_0 and $\tilde{s}_0$ iff there exist local coordinates $s^1, \ldots, s^p$ about s_0 in S and $\tilde{s}^1, \ldots, \tilde{s}^p$ about $\tilde{s}_0$ in $\tilde{S}$ such that the kth order Taylor polynomial of $f^j(s^1, \ldots, s^p)$ about s_0 equals the kth order Taylor polynomial of $\tilde{f}^j(\tilde{s}^1, \ldots, \tilde{s}^p)$ about $\tilde{s}_0$, for $j = 1, \ldots, m$.

Consequently, if f and $\tilde{f}$ are of class C^ω, and $k = \infty$ --i.e., f and $\tilde{f}$ have contact of at least order k at s_0 and $\tilde{s}_0$, for every positive integer k --then $f = \tilde{f} \circ F$, where $F\colon S \to \tilde{S}$ is the (local) diffeomorphism defined by $\tilde{s}^i = s^i$, $i = 1, \ldots, p$. In particular $f(S) = \tilde{f}(\tilde{S})$.

In the study of submanifolds of a homogeneous space G/G_0, one often does not distinguish between a given submanifold and G-translates of it. Thus we extend the definition of contact as follows.

Let G be a group of diffeomorphisms on M. We say submanifolds $f: S \to M$ and $\tilde{f}: \tilde{S} \to M$ have G-contact of at least order k at $s \in S$ and $\tilde{s} \in \tilde{S}$ if there exists a transformation $A \in G$ such that f and $A \circ f$ have contact of at least order k at s and $\tilde{s}$.

Definition. Imbedded submanifolds $f: S \to M$ and $\tilde{f}: \tilde{S} \to \tilde{M}$ are G-congruent if there exists a transformation $A \in G$ such that $A\, f(S) = \tilde{f}(\tilde{S})$.

Since f and $\tilde{f}$ are imbeddings, it is clear that they are G-congruent iff there is an $A \in G$ and a diffeomorphism $F: S \to \tilde{S}$ such that $A \circ f = \tilde{f} \circ F: S \to M$.

I.2. The Grassmann bundle. Let $\pi: L(M) \to M$ denote the principal $GL(m;\mathbb{R})$-bundle of linear frames on M. Then the Grassmann bundle is the associated fibre bundle $G_{m,p}(M) = L(M) \times_{GL(m;\mathbb{R})} G_{m,p}$, where $G_{m,p}$ is the Grassmann manifold of all p-dimensional subspaces of $\mathbb{R}^m$ on which $GL(m;\mathbb{R})$ acts on the left in the usual way.

Let $f: S \to M$ be a p-dimensional submanifold, and let $L_0 = f^{-1}L(M)$ be the pull-back of $L(M)$ to S. There is a naturally defined smooth map $\lambda_0: L_0 \to G_{m,p}$ given by $\lambda_0(u) \equiv u_f = u^{-1}f_*S_s$ where $f(s) = \pi(u)$. This map is equivariant with respect to the right action of $GL(m;\mathbb{R})$ on L_0 and the left action of $GL(m;\mathbb{R})$ on $G_{m,p}$, namely

$\lambda_0(uK) = K^{-1}\lambda_0(u)$, where $u \in L_0$, $K \in GL(m;\mathbb{R})$.

A <u>local frame field along</u> f is a smooth local cross section of $L_0 \to S$. Any such local frame field u along f gives a local representation of the tangent map $T_f: S \to G_{m,p}(M) = L(M) \times_{GL(m;\mathbb{R})} G_{m,p}$, namely $T_f = [u, \lambda_0 \circ u]$, where $[u, x]$ denotes the equivalence class of $(u, x) \in L(M) \times G_{m,p}$ with respect to the action of $GL(m;\mathbb{R})$.

If $A: M \to M$ is a diffeomorphism and $u \in L_0$ is a frame of f at $s \in S$, then A_*u is a frame of $A \circ f: S \to M$ at s and $(A_*u)_{A\circ f} = u_f$.

I.3. <u>Adapted frames of a homogeneous space</u>. Let G be a transitive Lie transformation group on M and fix an origin $0 \in M$. If G_0 denotes the isotropy subgroup of G at 0 , i.e. $G_0 = \{A \in G: A(0) = 0\}$, then the map $\pi: G \to M$, $\pi(A) = A(0)$, induces a diffeomorphism $G/G_0 \cong M$.

Choosing a basis $e_1, \ldots, e_m$ of the tangent space M_0 gives rise to a natural bundle map $h_0: G \to L(M)$ defined by $h_0(A) = A_*(e_1, \ldots, e_m)$. It is a bundle map because if $K \in G_0$, and if $\rho_0: G_0 \to GL(m, \mathbb{R})$ is the linear isotropy representation of G_0 with respect to the basis $e_1, \ldots, e_m$, then for any $A \in G$ we have $h_0(AK) = h_0(A)\rho_0(K)$. The subbundle $h_0(G)$ is called the set of <u>adapted frames</u> on M .

Let $\mathfrak{g}$ and $\mathfrak{g}_0$ denote the Lie algebras of G and G_0 , respectively. Choose a vector subspace $\mathfrak{m}_0$ of $\mathfrak{g}$ complementary to $\mathfrak{g}_0$, (if G_0 is compact we choose $\mathfrak{m}_0$ to be $Ad(G_0)$-invariant), and choose a basis $E_1, \ldots, E_m$ of $\mathfrak{m}_0$ such that $\pi_* E_i = e_i$, $i = 1, \ldots, m$. If $K \in G_0$ and $\rho_0(K) = (K^i_j)_{1 \leq i,j \leq m}$, then $\mathrm{Ad}(K)E_j = \sum_{i=1}^{m} K^i_j E_i \bmod \mathfrak{g}_0$. We use this fact to compute $\rho_0(K)$.

I.4. Zeroth order frames of a submanifold. Let $f: S \to M$ be a connected, smoothly imbedded, p-dimensional submanifold of the homogeneous space $M = G/G_0$. Given $s_0 \in S$, the frame construction is local, i.e., takes place on a neighborhood of s_0. Thus, to simplify the exposition, we shrink S to a neighborhood of s_0 whenever necessary in order to assume all maps defined on all of S.

Definition. A zeroth order frame at $s \in S$ is an element $A \in G$ such that $\pi(A) = f(s)$. (By calling A a frame we are tacitly identifying A with $h_0(A)$).

Let L_0 denote the set of all zeroth order frames on S. $L_0 = \{A \in G: \pi(A) \in f(S)\}$, which is just the pull-back by f of the bundle $\pi: G \to M$. It is a principal G_0-bundle over S, and a smooth submanifold of G.

A zeroth order frame field along f is a smooth cross section of $L_0 \to S$, i.e., is a smooth map $u: S \to G$ such that $\pi \circ u = f$.

If $u: S \to G$ is a zeroth order frame field along f, then any other is given by $\tilde{u} = uK$, where $K: S \to G_0$ is a smooth map.

Zeroth order frames A and B at $s \in S$ have the same orientation if $B = AK$ for some K in the connected component of the identity of G_0, that is, A and B are in the same connected component of the fibre of L_0 over s.

The natural map λ_0 of I.2 restricts to L_0. Explicitly, if $A \in L_0 \subseteq G$ lies over $s \in S$, i.e., $f(s) = \pi(A)$, then $\lambda_0(A) \equiv A_f \equiv (h_0 A)^{-1} f_* S_s$ which is a p-dimensional subspace of $\mathbb{R}^m$. This map is G_0-equivariant: if $A \in L_0$, $K \in G_0$ then $\lambda_0(AK) = \rho_0(K)^{-1}\lambda_0(A)$. For any $B \in G$, the zeroth order frames of $\tilde{f} \equiv B \circ f: S \to M$ are just BL_0 and the associated map $\tilde{\lambda}_0: BL_0 \to G_{m_0,p}$ is given by $\tilde{\lambda}_0(BA) = \lambda_0(A)$, for any $A \in L_0$.

I.5. Transformation groups. Let K be a Lie transformation group on a smooth manifold X. We define a local cross section of K on X to be a smoothly imbedded submanifold W of X such that W meets each orbit of K at most once, and those orbits it meets it crosses transversally.

For any point $x \in X$, we denote the isotropy subgroup of K at x by K_x, that is $K_x = \{A \in K: Ax = x\}$. A closed subgroup of K is called an isotropy subgroup if it is the isotropy subgroup of K at some point of X. The isotropy subgroups at points of a single orbit are all K-conjugate. An orbit is said to be of type K/H, where H is a closed subgroup of K, if the isotropy subgroup at any point of the orbit is K-conjugate to H. Finally, we let $X_{(H)}$ denote the set of all points in X lying on orbits of type K/H.

Theorem. Suppose K is compact, and H is an isotropy subgroup of K. Then $K_{(H)}$ is a smoothly imbedded submanifold of X and is a smooth K/H-bundle over its orbit space $X^*_{(H)} \equiv H\backslash X_{(H)}$. Furthermore, if

$x \in K_{(H)}$, and if we replace H by a K-conjugate subgroup (if necessary) so that we may assume $H = K_x$, then there is a local cross section $W \subseteq X_{(H)}$ passing through x and such that $K_y = H$ for every $y \in W$. The map $K/H \times W \to K \cdot W$ given by $(A \cdot H, y) \to Ay$ is an equivariant diffeomorphism.

Proof. We refer to Bredon [B], p. 309.//

Let $g: Y \to X$ be a smooth mapping from a smooth manifold Y . We say g is of constant type if there is an isotropy subgroup H of K such that $g(Y) \subseteq X_{(H)}$.

Corollary. Suppose K is compact and $g: Y \to X$ is of constant type, say $g(Y) \subseteq X_{(H)}$, where H is some isotropy subgroup of K . Then for any point $y_0 \in Y$, there exists an open neighborhood Y' of y_0 in Y and a local cross section $W \subseteq X_{(H)}$ such that (1) $g(y_0) \equiv x_0 \in W$ and (2) there exists a smooth map $A: Y' \to K$ such that $\tilde{g}(y) \equiv A(y)g(y) \in W$ for every $y \in Y'$.

I.6. First order frames. Suppose W_1 is a local cross section of the action (G_0, ρ_0) on $G_{m_0,\rho}$. We say that f has the type of W_1 if there exists a zeroth order frame field u along f such that $\lambda_0 \circ u(S) \subseteq W_1$. If f has the type of W_1 , we let $L_1 = \lambda_0^{-1} W_1$ and call L_1 the set of first order frames on f (with respect to W_1) . Then L_1 is a smooth submanifold of L_0 since W_1 cuts orbits transversally, so the inverse function theorem applies to λ_0 ; and $L_1 \to S$ is a trivial fibration as assured by the above definition of type.

We define a first order frame field along f to be a smooth cross section of $L_1 \to S$. Two first order frames over $s \in S$ are said to have the same orientations if they are in the same connected component of the fibre of L_1 over s .

The fibre of L_1 over $s \in S$ is $A \cdot G_1(x)$, where $A \in L_1$, $\pi(A) = f(s)$, $\lambda_0(A) = x \in W_1$, and $G_1(x)$ is the isotropy subgroup of G_0 at x , i.e., $G_1(x) = \{K \in G_0 : \rho_0(K)x = x\}$. By the equivariance of λ_0 and the definition of cross section of ρ_0 , it follows that λ_0 is constant on the fibres of L_1 . Hence the smooth map $\lambda_0 \circ u : S \to W_1$ does not depend on the choice of first order frame field u along f .

Choose a coordinate system $x^1, \ldots, x^{\mu_1}$ on W_1 (shrinking W_1 and S to an open submanifold, if necessary). We call the functions $k^i = x^i \circ \lambda_0 \circ u$, $i = 1, \ldots, {}_1$ the first order invariants of f , where u is any first order frame field along f . For any $s \in S$, the numbers $k^i(s)$, $i = 1, \ldots, \mu_1$ are called the first order invariants of f at s .

<u>Theorem 1</u>. Let $f : S \to M$ and $\tilde{f} : \tilde{S} \to M$ be smoothly imbedded p-dimensional submanifolds on which first order frames can be constructed. (Use the above notation for f , and the same with tildes for $\tilde{f}$) . Then f and $\tilde{f}$ have at least first order G-contact at $s \in S$ and

$\tilde{s} \in \tilde{S}$ iff they are both the type of a local cross section W_1 of ρ_0 and they have the same first order invariants at s and $\tilde{s}$, respectively. Furthermore, the element $\tilde{A}A^{-1} \in G$ will transform f into at least first order contact with $\tilde{f}$ at $\tilde{f}(\tilde{s})$, where A and $\tilde{A}$ are any first order frames f and $\tilde{f}$ at s and $\tilde{s}$, respectively.

Proof. Suppose f and $\tilde{f}$ have at least first order G-contact at $s \in S$ and $\tilde{s} \in \tilde{S}$. Then there exists $B \in G$ such that $B \circ f$ and $\tilde{f}$ have at least first order contact at s and $\tilde{s}$; i.e., $T_{B\circ f}(s) = T_{\tilde{f}}(\tilde{s})$. Let A and $\tilde{A}$ be zeroth order frames at s and $\tilde{s}$, respectively. Then BA is a zeroth order frame at s of $B \circ f$ and $T_{B\circ f}(s) = [BA, \lambda_0(A)]$, while $T_{\tilde{f}}(\tilde{s}) = [\tilde{A}, \tilde{\lambda}_0(\tilde{A})]$. Since these are equal, there exists $K \in G_0$ such that $BAK^{-1} = \tilde{A}$ and $\rho_0(K)\lambda_0(A) = \tilde{\lambda}_0(\tilde{A})$. But the latter equation says that $\lambda_0(A)$ and $\tilde{\lambda}_0(\tilde{A})$ lie in the same G_0-orbit, and thus a common local cross section W_1 can be chosen for f and $\tilde{f}$. Then, choosing A and $\tilde{A}$ above to be first order frames at s and $\tilde{s}$ of f and $\tilde{f}$, respectively, it follows that $\lambda_0(A) = \tilde{\lambda}_0(A)$, since they both lie in W_1 and both lie on the same G_0-orbit.

Conversely, suppose f and $\tilde{f}$ are both of type W_1 and $\lambda_0(A) = \tilde{\lambda}_0(\tilde{A})$ for any first order frames A and $\tilde{A}$ at s and $\tilde{s}$ of f and $\tilde{f}$, respectively. Let $B = \tilde{A}A^{-1} \in G$. Then $T_{B\circ f}(s) = [BA, \lambda_0(A)] = [\tilde{A}, \tilde{\lambda}_0(\tilde{A})] = T_{\tilde{f}}(\tilde{s})$, i.e., f and $\tilde{f}$ have at least first order G-contact at s and $\tilde{s}$. //

Suppose now that G_0 is compact. We say that $f: S \to M$ is of constant type at zeroth order if $\lambda_0: L_0 \to G_{m_0,p}$ is of constant type with respect to the action (G_0, ρ_0) on $G_{m_0,p}$.

<u>Proposition</u>. If G_0 is compact and if f is of constant type at zeroth order, say $\lambda_0(L_0) \subseteq G_{m_0,p(G_1)}$ where G_1 is an isotropy subgroup of G_0 , then for any point $s_0 \in S$ there is an open neighborhood S' of s_0 in S and a local cross section $W_1 \subseteq G_{m_0,p(G_1)}$ such that $f|_{S'}$ is the type of W_1 and the isotropy subgroup of G_0 at any point of W_1 is G_1 .

I.7. <u>Second order frames</u>. We have now constructed the first order frames $L_1 \to S$, using the local cross section W_1 of the action (G_0, ρ_0) on $G_{m_0,p}$, namely $L_1 = \lambda_0^{-1}W_1$, where $\lambda_0: L_0 \to S$. The fibres of L_1 are isomorphic to isotropy subgroups of G_0 at points of W_1 . If G_0 is not compact, these subgroups form a family of closed subgroups of G_0 smoothly parametrized by W_1 .

To make this explicit, let $G_1(x) = \{K \in G_0: \rho_0(K)x = x\}$, where $x \in W_1$. If $\phi: G_0 \times W_1 \to G_{m_0,p}$ is the smooth map defined by $\phi(K, x) = \rho_0(K)x$, then for each $x \in W_1$, $G_1(x) \times \{x\} = \phi^{-1}\{x\}$. In particular the dimension of $G_1(x)$ does not depend on $x \in W_1$, (on an open subset).

Let $\mathfrak{g}_1(x)$ be the Lie algebra of $G_1(x)$, choose a complementary vector subspace $\mathfrak{m}_1(x)$ to $\mathfrak{g}_1(x)$ in $\mathfrak{g}_0$, and choose a basis

$E_{m_0+1}(x), \ldots, E_{m_1}(x)$ of $\mathfrak{m}_1(x)$, where $m_1 - m_0 = \dim \mathfrak{m}_1(x)$, for $x \in W_1$. We may assume these choices depend smoothly on $x \in W_1$ in the sense that the maps $W_1 \to \mathfrak{g}$ given by $x \to E_i(x)$ are smooth, $i = m_0 + 1, \ldots, m_1$. If $G_1(x)$ is a fixed compact subgroup G_1 for all $x \in W_1$, we choose $\mathfrak{m}_1$ to be $\mathrm{ad}(G_1)$-invariant.

Let $M^{(1)} = \bigcup_{x \in W_1} G/_{G_1(x)} \times \{x\}$, which by an abuse of notation we denote $G/_{G_1} \times W_1$. (Note there is no abuse if G_0 is compact.) There is a natural inclusion $M^{(1)} \hookrightarrow G_{m_0,p}(M) = G \times_{G_0} G_{m_0,p}$ given by $(A \cdot G_1(x), x) \mapsto [A, x]$. We assume $M^{(1)}$ is a smoothly imbedded submanifold of $G_{m_0,p}(M)$. Since $T_f^{(1)} \equiv T_f \colon S \to G_{m_0,p}(M)$ is given by $T_f^{(1)} = [u, \lambda_0 \circ u]$, where u is any first order frame field along f, it follows that $T_f^{(1)}(S) \subset M^{(1)}$ and $T_f^{(1)} = (u \cdot G_1(\lambda_0 \circ u), \lambda_0 \circ u)$.

Define $h_1 \colon G \times W_1 \to L(M^{(1)})$ by $h_1(A, x)$

$$= (A_* \pi_1(x)_*(E_1, \ldots, E_{m_0}, E_{m_0+1}(x), \ldots, E_m(x)), \frac{\partial}{\partial x^1}, \ldots, \frac{\partial}{\partial x^{\mu_1}}),$$

where $A \in G$, $x \in W_1$, $\pi_1(x) \colon G \to G/G_1(x)$ is the natural projection, and $x^1, \ldots, x^{\mu_1}$ is the chosen coordinate system on W_1. This is a linear frame on $M^{(1)}$ at $(\pi_1(x)A, x)$.

For each $x \in W_1$, let $\rho_1 \colon G_1(x) \to GL(m_1 + \mu_1; \mathbb{R})$ denote the linear representation of $G_1(x)$ on $\mathbb{R}^{m_1+\mu_1} = \mathbb{R}^{m_1} \times \mathbb{R}^{\mu_1}$ given on $\mathbb{R}^{m_1}$ by the adjoint action of $G_1(x)$ on $\mathfrak{m}_0 + \mathfrak{m}_1(x)$ modulo $\mathfrak{g}_1(x)$ with respect to the basis $E_1, \ldots, E_{m_0}, E_{m_0+1}(x), \ldots, E_{m_1}(x)$; and on $\mathbb{R}^{\mu_1}$ by the identity transformation.

Define the smooth map $\lambda_1 : L_1 \to G_{m_1+\mu_1,p}$ by $\lambda_1(A)$ $= h_1(A, \lambda_0(A))^{-1}(T_f^{(1)})_* S_s$ = p-plane in $\mathbb{R}^{m_1+\mu_1}$, where $\pi(A) = f(s)$. Then λ_1 is equivariant, i.e., if $A \in L_1$, $K \in G_1(\lambda_0(A))$, then $\lambda_1(AK) = \rho_1(K^{-1})\lambda_1(A)$.

Let $\psi_1 : G_{m_1+\mu_1,p} \to G_{m_0,p}$ (defined on an open dense subset) be the smooth map induced by the projection map $\mathbb{R}^{m_1+\mu_1} = \mathbb{R}^{m_0} \times \mathbb{R}^{m_1-m_0} \times \mathbb{R}^{\mu_1} \to \mathbb{R}^{m_0}$. Then for any $x \in W_1$ and any $K \in G_1(x)$, $\psi_1 \circ \rho_1(K)$ $= \rho_0(K) \circ \psi_1 : G_{m_1+\mu_1,p} \to G_{m_0,p}$; and $\psi_1 \circ \lambda_1 : L_1 \to G_{m_0,p}$ is the restriction of λ_0 to L_1 , i.e., $\psi_1 \circ \lambda_1 = \lambda_0|_{L_1}$. Since $\psi_1\rho_1 = \rho_0\psi_1$, it follows that for any $x \in W_1$, $\rho_1(G_1(x)) \cdot \psi_1^{-1}\{x\} \subseteq \psi_1^{-1}\{x\}$.

We define a local cross section of ρ_1 to be a smoothly imbedded submanifold W_2 of $\psi_1^{-1}W_1$ in $G_{m_1+\mu_1,p}$ such that: (1) for each $x \in W_1$, $W_2 \cap \psi_1^{-1}\{x\}$ is a local cross section (as in I.5) of the action $(G_1(x), \rho_1)$ on $\psi_1^{-1}\{x\}$; and (2) $\psi_1 : W_2 \to W_1$ is a smooth submersion onto an open submanifold of W_1 .

We say that f has the type of a local cross section W_2 of ρ_1 if there exists a first order frame field u along f such that $\lambda_1 \circ u(S) \subseteq W_2$. If f has the type of W_2 , we let $L_2 = \lambda_1^{-1}W_2$, and call L_2 the set of second order frames on f (with respect to W_2) . Then L_2 is a smooth submanifold of L_1 since W_2 cuts orbits transversally, so the inverse function theorem applies to λ_1 ; and $L_2 \to S$ is a trivial fibration as assured by the above definition of type.

We define a second order frame field along f to be a smooth cross section of $L_2 \to S$. Two second order frames A and B over $s \in S$ are said to have the same orientation if they are in the same connected component of the fibre of L_2 over s.

The fibre of L_2 over $s \in S$ is $A \cdot G_2(y)$, where $A \in L_2$, $\pi(A) = f(s)$, $\lambda_1(A) = y \in W_2$, and $G_2(y)$ is the isotropy subgroup of $G_1(\psi_1 y)$ at y, i.e. $G_2(y) = \{K \in G_1(\psi_1 y): \rho_1(K)y = y\}$. By the equivariance of λ_1 and the definition of cross section of ρ_1, it follows that λ_1 is constant on the fibres of L_2. Hence the smooth map $\lambda_1 \circ u: S \to W_2$ does not depend on the choice of second order frame field u along f.

Let $x^1, \ldots, x^{\mu_1}$ be the coordinate system on W_1 chosen to define the first order invariants $k^1, \ldots, k^{\mu_1}$ of f. Since $\psi_1: W_2 \to W_1$ is a submersion, the functions $x^1 \circ \psi_1, \ldots, x^{\mu_1} \circ \psi_1$ are independent on W_2. Continuing to denote $x^i \circ \psi_1$ by x^i, $i = 1, \ldots, \mu_1$, we extend these functions to a coordinate system $x^1, \ldots, x^{\mu_2}$ of W_2, $\mu_2 = \dim W_2 \geq \mu_1$. We call $k^i = x^i \circ \lambda_1 \circ u$, $i = \mu_1 + 1, \ldots, \mu_2$ the second order invariants of f, where u is any second order frame field along f.

The proof of the next theorem is similar to that of the theorem in section I.6.

<u>Theorem.</u> Let $f: S \to M$ and $\tilde{f}: \tilde{S} \to M$ be smoothly imbedded p-dimensional submanifolds on which first and second order frames can be constructed. (Use the above notation for f, and the same with tildes

for $\tilde{f}$) . Then f and $\tilde{f}$ have at least second order G-contact at $s \in S$ and $\tilde{s} \in \tilde{S}$ iff they are both the type of a local cross section W_2 of ρ_1 and they have the same first and second order invariants at s and $\tilde{s}$, respectively, i.e. $\lambda_1(A) = \tilde{\lambda}_1(\tilde{A})$ where A and $\tilde{A}$ are any second order frames of f and $\tilde{f}$ at s and $\tilde{s}$, respectively. Furthermore, the element $\tilde{A}A^{-1} \in G$ will transform f into at least second order contact with $\tilde{f}$ at $\tilde{f}(\tilde{s})$.

Suppose now that G_0 is compact, or that $G_1(x)$ equals a fixed compact subgroup G_1 of G_0 for every $x \in W_1$. We say then that $f: S \to M$ is of constant type at first order if $\lambda_1: L_1 \to G_{m_1+\mu_1,p}$ is of constant type with respect to the action (G_1, ρ_1) .

<u>Proposition</u>. If f is of constant type at first order, say $\lambda_1(L_1) \subseteq G_{m_1+\mu_1,p(G_2)}$, where G_2 is an isotropy subgroup of G_1 , then for any point $s_0 \in S$ there is an open neighborhood S' of s_0 in S and a local cross section $W_2 \subseteq \psi_1^{-1}W_1 \cap G_{m_1+\mu_1,p(G_2)}$ such that (i) $f|_{S'}$ is the type of W_2 , (ii) the isotropy subgroup of G_1 at any point of W_2 is G_2 , and (iii) $\psi_1: W_2 \to W_1$ is a submersion onto an open submanifold of W_1 .

<u>Proof</u>. This follows from the Corollary in section I.5. For a more explicit expression of the action ρ_1 we refer to section I.10.//

I.8. Third and higher order frames. We shall describe the construction of the third order frames from the second order frames. Since this step involves every feature present in the step from order k to $k+1$ for any $k \geq 1$, we shall let it serve as the model for that general step.

Suppose we have already constructed the first and second order frames on $f: S \to M$. That is, f is of type W_1 and of type W_2, where $W_1 \subseteq G_{m_0,p}$ is a local cross section of ρ_0 and $W_2 \subseteq \psi_1^{-1} W_1 \subseteq G_{m_1+\mu_1,p}$ is a local cross section of ρ_1. We have the maps: $\lambda_0: L_0 \to G_{m_p,p}$ which gives $L_1 = \lambda_0^{-1} W_1$, the first order frames on f; and $\lambda_1: L_1 \to G_{m_1+\mu_1,p}$ which gives $L_2 = \lambda_1^{-1} W_2$, the second order frames on f. The isotropy subgroup of G_0 at $x \in W_1$ is $G_1(x)$, they have Lie algebras $\mathfrak{g}_0$ and $\mathfrak{g}_1(x)$, respectively, $\mathfrak{g}_0 = \mathfrak{g}_1(x) \oplus \mathfrak{m}_1(x)$. We have chosen bases $E_1, \ldots, E_{m_0}$ of $\mathfrak{m}_0$, where $\mathfrak{g} = \mathfrak{g}_0 \oplus \mathfrak{m}_0$, and $E_{m_0+1}(x), \ldots, E_{m_1}(x)$ of $\mathfrak{m}_1(x)$.

For $y \in W_2$, let $x = \psi_1 y \in W_1$, let $G_2(y) = \{K \in G_1(x): \rho_1(K)y = y\}$ = a closed subgroup of $G_1(x)$, let $\mathfrak{g}_2(y)$ be its Lie algebra, choose a complementary vector subspace $\mathfrak{m}_2(y)$ to $\mathfrak{g}_2(y)$ in $\mathfrak{g}_1(x)$, and choose a basis $E_{m_1+1}(y), \ldots, E_{m_2}(y)$ of $\mathfrak{m}_2(y)$, where $m_2 - m_1 = \dim \mathfrak{m}_1(y)$ is independent of $y \in W_2$. We assume these subspaces and subalgebras vary smoothly with y in the sense that the maps $W_2 \to \mathfrak{g}$ given by $y \mapsto E_i(y)$ are smooth, $i = m_1 + 1, \ldots, m_2$. In the event that $G_2(y)$ is a fixed compact subgroup G_2 for every $y \in W_2$, we choose $\mathfrak{m}_2$ to be $\mathrm{ad}(G_2)$-invariant.

Let $M^{(2)} = \bigcup_{y \in W_2} G/G_2(y) \times \{y\}$, which by an abuse of notation we denote by $G/G_2 \times W_2$. (Note there is no abuse if $G_2(y) = G_2$ does not depend on $y \in W_2$) . Using $h_1 \colon G \times W_1 \to L(M^{(1)})$, we may represent $G_{m_1+\mu_1,p}(M^{(1)})$ as $(G \times W_1) \times_{G_1} G_{m_1+\mu_1,p} \overset{\text{def}}{=}$ the space of equivalence classes of the equivalence relation on $G \times W_1 \times G_{m_1+\mu_1,p}$ defined by: $((A, x), y) \sim ((A', x'), y')$ if $x' = x$ and $A' = AK^{-1}$, $y' = \rho_1(K)y$ for some $K \in G_1(x)$. Then there is an inclusion $M^{(2)} \hookrightarrow G_{m_1+\mu_1,p}(M^{(1)})$ $= (G \times W_1) \times_{G_1} G_{m_1+\mu_1,p}$ given by $(A \cdot G_2(y), y) \to [(A, \psi_1 y), y]$, where $A \in G$, $y \in W_2$, and $[(A, \psi_1 y), y]$ denotes the equivalence class of $(A, \psi_1 y, y) \in G \times W_1 \times G_{m_1+\mu_1,p}$. We assume that $M^{(2)}$ is a smoothly imbedded submanifold of $G_{m_1+\mu_1,p}(M^{(1)})$. (Always the case if G_0 is compact).

Since $T_f^{(2)} \colon S \to G_{m_1+\mu_1,p}(M^{(1)})$ is given by $T_f^{(2)}$ $= [(u, \lambda_0 \circ u), \lambda_1 \circ u] \in (G \times W_1) \times_{G_1} G_{m_1+\mu_1,p}$, where u is any second order frame field along f , it follows that $T_f^{(2)}(S) \subseteq M^{(2)}$ and is represented by $T_f^{(2)} = (\pi_2(\lambda_1 \circ u) \circ u, \lambda_1 \circ u)$, where for $y \in W_2$, $\pi_2(y) \colon G \to G/G_2(y)$ is the natural projection.

Define $h_2 \colon G \times W_2 \to L(M^{(2)})$ by $h_2(A, y)$

$$= (A_* \pi_2(y)_* (E_1, \ldots, E_{m_0}, E_{m_0+1}(x), \ldots, E_{m_1}(x), E_{m_1+1}(y), \ldots, E_{m_2}(y)), \frac{\partial}{\partial x^1}, \ldots, \frac{\partial}{\partial x^{\mu_2}}) ,$$

where $A \in G$, $y \in W_2$, $x = \psi_1 y \in W_1$ and $x^1, \ldots, x^{\mu_2}$ is the chosen coordinate system on W_2 . Clearly $h_2(A, y)$ is a linear frame on $M^{(2)}$ at $(\pi_2(y)A, y)$.

For each $y \in W_2$, let $\rho_2 \colon G_2(y) \to GL(m_2 + \mu_2; \mathbb{R})$ denote the linear representation of $G_2(y)$ on $\mathbb{R}^{m_2+\mu_2} = \mathbb{R}^{m_2} \times \mathbb{R}^{\mu_2}$ given on $\mathbb{R}^{m_2}$ by the adjoint action of $G_2(y)$ on $\mathcal{M}_0 + \mathcal{M}_1(\psi_1 y) + \mathcal{M}_2(y)$ modulo $\mathcal{G}_2(y)$ with respect to the basis $E_1, \ldots, E_{m_2}$; and on $\mathbb{R}^{\mu_2}$ by the identity transformation.

Define the smooth map $\lambda_2 \colon L_2 \to G_{m_2+\mu_2,p}$ by $\lambda_2(A) = h_2(A, \lambda_1(A))^{-1}(T_f^{(2)})_* S_s$ = p-plane in $\mathbb{R}^{m_2+\mu_2}$, where $\pi(A) = f(s)$. Then λ_2 is equivariant; i.e., if $A \in L_2$ and $K \in G_2(\lambda_1(A))$, then $\lambda_2(AK) = \rho_2(K^{-1})\lambda_2(A)$. Thus λ_2 maps a fibre of L_2 onto a ρ_2-orbit in $G_{m_2+\mu_2,p}$.

Let $\psi_2 \colon G_{m_2+\mu_2,p} \to G_{m_1+\mu_1,p}$ (defined on an open dense subset) be the rational map induced by the projection $\mathbb{R}^{m_2+\mu_2} = \mathbb{R}^{m_1} \times \mathbb{R}^{m_2-m_1} \times \mathbb{R}^{\mu_1} \times \mathbb{R}^{\mu_2-\mu_1} \to \mathbb{R}^{m_1} \times \mathbb{R}^{\mu_1} = \mathbb{R}^{m_1+\mu_1}$. Then for any $y \in W_2$ and any $K \in G_2(y)$, we have $\psi_2 \circ \rho_2(K) = \rho_1(K) \circ \psi_2 \colon G_{m_2+\mu_2,p} \to G_{m_1+\mu_1,p}$. Furthermore, $\psi_2 \circ \lambda_2 \colon L_2 \to G_{m_1+\mu_1,p}$ is the restriction of λ_1 to L_2, i.e., $\psi_2 \circ \lambda_2 = \lambda_1|_{L_2}$. Since $\psi_2\rho_2 = \rho_1\psi_2$, it follows that for any $y \in W_2$, $\rho_2(G_2(y)) \cdot \psi_2^{-1}\{y\} \subseteq \psi_2^{-1}\{y\}$.

We define a local cross section of ρ_2 to be a smoothly imbedded submanifold W_3 of $\psi_2^{-1} W_2$ in $G_{m_2+\mu_2,p}$ such that: (1) for each $y \in W_2$, $W_2 \cap \psi_2^{-1}\{y\}$ is a local cross section (as in I.5) of the action $(G_2(y), \rho_2)$ on $\psi_2^{-1}\{y\}$; and (2) $\psi_2 \colon W_3 \to W_2$ is a smooth submersion onto an open submanifold of W_2.

We say that f has the type of a local cross section W_3 of ρ_2 if there exists a second order frame field u along f such that $\lambda_2 \circ u(S) \subseteq W_3$. If f has the type of W_3, we let $L_3 = \lambda_2^{-1} W_3$, and call L_3 the set of third order frames on f (with respect to W_3). Then L_3 is a smooth submanifold of L_2 since W_3 cuts orbits transversally, so the inverse function theorem applies to λ_2; and $L_3 \to S$ is a trivial fibration as assured by the above definition of type.

We define a third order frame field along f to be a smooth cross section of $L_3 \to S$. Two third order frames over $s \in S$ are said to have the same orientation if they are in the same connected component of the fibre of L_3 over s.

The fibre of L_3 over $s \in S$ is $A \cdot G_3(z)$, where $A \in L_2$, $\pi(A) = f(s)$, $\lambda_2(A) = z \in W_3$, and $G_3(z)$ is the isotropy subgroup of $G_2(\psi_2 z)$ at z, i.e., $G_3(z) = \{K \in G_2(\psi_2 z) \colon \rho_2(K)z = z\}$. By the equivariance of λ_2 and the definition of cross section of ρ_2, it follows that λ_2 is constant on the fibres of L_3. Hence the smooth map $\lambda_2 \circ u \colon S \to W_3$ does not depend on the choice of third order frame field u along f.

Let $x^1, \ldots, x^{\mu_2}$ be the coordinate system on W_2 chosen to define h_2 and the second order invariants $k^1, \ldots, k^{\mu_2}$ of f. Since $\psi_2 \colon W_3 \to W_2$ is a submersion, the functions $x^i \circ \psi_2$, $i = 1, \ldots, \mu_2$ are independent on W_3. Continuing to denote $x^i \circ \psi_2$ by x^i on W_3, $i = 1, \ldots, \mu_2$, we extend this set of functions to a coordinate system $x^1, \ldots, x^{\mu_3}$ of W_3, $\mu_3 = \dim W_3 \geqslant \mu_2$. We call the functions

$k^i \equiv x^i \circ \lambda_2 \circ u$, $i = \mu_2 + 1, \ldots, \mu_3$ the third order invariants of f , where u is any third order frame field along f . For any $s \in S$, the numbers $k^i(s)$, $i = \mu_2 + 1, \ldots, \mu_3$ are called the third order invariants of f at s .

This completes the construction at the third order. Since this has really defined the construction from order k to $k + 1$ for any $k \geqslant 1$, we state the following theorem in the general case. The construction itself should make the proof quite transparent.

Theorem. Let $f: S \to M$ and $\tilde{f}: \tilde{S} \to M$ be smoothly imbedded p-dimensional submanifolds on which j^{th} order frames can be constructed, for $j = 0, 1, \ldots, k$, where k is a positive integer. (Use the above notation for f and the same with tildes for $\tilde{f}$). Then f and $\tilde{f}$ have at least k^{th} order G-contact at $s \in S$ and $\tilde{s} \in \tilde{S}$ iff they are both the type of a local cross section W_k of ρ_{k-1} , and they have the same j^{th} order invariants at s and $\tilde{s}$, respectively, for $j = 1, \ldots, k$. That is, $\lambda_{k-1}(A) = \tilde{\lambda}_{k-1}(\tilde{A})$ where A and $\tilde{A}$ are any k^{th} order frames of f and $\tilde{f}$ at s and $\tilde{s}$, respectively. Furthermore, in the latter case, $\tilde{A}A^{-1} \in G$ transforms f into at least k^{th} order contact with $\tilde{f}$ at $\tilde{f}(\tilde{s})$.

Suppose now that G_0 is compact, or that $G_k(x)$ equals a fixed compact subgroup G_k of G_0 for every $x \in W_k$. We say then that $f: S \to M$ is of constant type at k^{th} order if $\lambda_k: L_k \to G_{m_k+\mu_k, p}$ is of constant type with respect to the action (G_k, ρ_k) .

Proposition. If f is of constant type at k^{th} order, say $\lambda_k(L_k) \subseteq G_{m_k+\mu_k,p}(G_{k+1})$, where G_{k+1} is an isotropy subgroup of G_k , then for any point $s_0 \in S$ there is an open neighborhood S' of s_0 in S and a local cross section $W_{k+1} \subseteq \psi_k^{-1}W_k \cap G_{m_k+\mu_k,p}(G_{k+1})$ such that (i) $f|S'$ is the type of W_{k+1} , (ii) the isotropy subgroup of G_k at any point of W_{k+1} is G_{k+1} , and (iii) $\psi_k\colon W_{k+1} \to W_k$ is a submersion onto an open submanifold of W_k .

I.9. Frenet frames. Let $f\colon S \to M$ be a smoothly imbedded p-dimensional submaniflod of the homogeneous space $M = G/G_0$, and suppose frames of all orders can be constructed on f .

For each integer $k \geqslant 1$, let n_k denote the number of independent invariants of order $\leqslant k$, i.e. $n_k = \dim \lambda_{k-1}(L_k) = \dim \lambda_{k-1} \circ u(S)$, where u is any k^{th} order frame field along f . Clearly $0 \leqslant n_1 \leqslant n_2 \leqslant \dots \leqslant p$. Moreover, the sequence $\dim L_0 \geqslant \dim L_1 \geqslant \dim L_2 \geqslant \dots \geqslant p$ eventually stabilizes. Thus there is a smallest integer $q \geqslant 1$ such that $n_k = n_q$ and $\dim L_k = \dim L_q$ for all $k \geqslant q$. Then the frames of order q are called the Frenet frames of f .

Part of the significance of the Frenet frames is contained in the following theorem, the proof of which is obvious. The conclusion of G-congruence in the real analytic case follows from the Theorem in section I.1. In section I.11 we prove the congruence theorem for the class C^{q+1} case.

<u>Theorem</u>. Let $f: S \to M$ and $\tilde{f}: \tilde{S} \to M$ be smoothly imbedded p-dimensional submanifolds of M. Suppose they are both the type of W_q, where q is the order of their Frenet frames, and suppose there is a one-to-one correspondence $\psi: S \to \tilde{S}$ such that f and $\tilde{f}$ have at least $(q+1)^{st}$ order G-contact at s and $\psi(s)$ for every $s \in S$. Then f and $\tilde{f}$ have at least k^{th} order G-contact at s and $\psi(s)$ for every positive integer k and for every $s \in S$. Hence, in the real analytic case, f and $\tilde{f}$ are G-congruent.

I.10. <u>The role of the Maurer-Cartan form of G</u>. The Maurer-Cartan form Ω of G is the $\mathfrak{g}$-valued left-invariant 1-form on G defined by $\Omega = L_{A^{-1}}{}^{*}dA$, where dA denotes the differential of the identity map on G and $L_{A^{-1}}$ is left-translation by A^{-1}, $A \in G$. The Maurer-Cartan equations are $d\Omega = -\frac{1}{2}[\Omega, \Omega]$.

Suppose that $f: S \to M = G/G_0$ is a p-dimensional smoothly imbedded submanifold. Let $L_0 \to S$ be the G_0-bundle of zeroth order frames on f.

The decomposition $\mathfrak{g} = \mathfrak{g}_0 + \mathfrak{m}_0$ decomposes Ω into $\omega_0 + \Theta_0$, where ω_0 is the $\mathfrak{g}_0$-component and Θ_0 is the $\mathfrak{m}_0$-component of Ω. Using the chosen basis $E_1, \ldots, E_{m_0}$ of $\mathfrak{m}_0$, we have $\Theta_0 = \sum_{i=1}^{m_0} \theta^i \otimes E_i$, where $\theta^1, \ldots, \theta^{m_0}$ are ordinary left-invariant 1-forms on G.

We make use of the following notation. For positive integers $n > p > 0$, let $\mathbb{R}^{n\times p}$ denote the space of all $n \times p$ real matrices, let $\mathbb{R}^{n\times p*}$ denote the open subset consisting of all $n \times p$ real matrices of rank p. Then $GL(p;\mathbb{R})$ acts on $\mathbb{R}^{n\times p*}$ by right multiplication and the quotient space is the real Grassmannian of p-planes in $\mathbb{R}^n$; $G_{n,p} = \mathbb{R}^{n\times p*}/GL(p;\mathbb{R})$. For any $X \in \mathbb{R}^{n\times p*}$ we denote its projection in $G_{n,p}$ by

$[X]$. On the open set $U = \left\{ [{A \atop Y}] \in G_{n,p} : A \in GL(p),\ Y \in \mathbb{R}^{(n-p)\times p} \right\}$ is a coordinate map $t\colon U \to \mathbb{R}^{(n-p)\times p}$ given by $t[{A \atop Y}] = YA^{-1}$. We shall refer to this as the standard coordinate system on $G_{n,p}$.

Lemma 1. Let $u\colon S \to G$ be a zeroth order frame field along f . Set $h \circ u = (e_1, \ldots, e_{m_0})$, where $e_i(s) = u(s)_* \pi_* E_i$, $i = 1, \ldots, m_0$ are vector fields along f . Let $\phi^1, \ldots, \phi^p$ be a coframe field on S , set $u^*\theta^i = \sum_{\alpha=1}^{p} x^i_\alpha \phi^\alpha$, $i = 1, \ldots, m_0$ for some smooth functions x^i_α on S , and set $X = (x^i_\alpha); S \to \mathbb{R}^{m_0 \times p}$. Then

(1) $f_*\colon S_s \to M_{f(0)}$ is given by $f_* = \sum_{i=1}^{m_0} (u^*\theta^i) \otimes e_i = \sum_{\alpha=1}^{p} \phi^\alpha \otimes \sum_{i=1}^{m_0} x^i_\alpha e_i$,

which implies that $X(s) \in \mathbb{R}^{m_0 \times p^*}$ for each $s \in S$;

(2) $\lambda_0 \circ u\colon S \to G_{m_0,p}$ is given by $\lambda_0 \circ u = [X]$.

Proof. (1) Let $v \in S_s$, then $f_* v = (\pi \circ u)_* v = \pi_{*u(s)}\, u_{*s}\, v$

$= \pi_{*u(s)}\, u(s)_{*1}\, u(s)^{-1}_{*u(s)}\, u_{*s} v = \pi_{*u(s)}\, u(s)_{*1} \Omega(u_{*s} v)$

$= u(s)_{*0}\, \pi_{*1}\, (u^*\Omega)(v) = u(s)_{*0}\, \pi_{*1}\, (u^*\omega_0 + u^*\Theta_0)(v) = u(s)_{*0}\, \pi_{*1} (u^*\Theta_0)(v)$

$= u(s)_{*0}\, \pi_{*1} \sum_{i=1}^{m_0} (u^*\theta^i)(v) \otimes E_i = \sum_{i=1}^{m_0} (u^*\theta^i)(v) \otimes u(s)_{*0}\, \pi_{*1}\, E_i$

$= \sum_{i=1}^{m_0} (u^*\theta^i)(v) \otimes e_i$, where 1 is the identity element of G , and $0 = \pi(1) \in M$.

(2) is an easy consequence of (1).//

From (1) of Lemma 1 and from the fact that f is an immersion it follows that p of the forms $u^*\theta^1, \ldots, u^*\theta^{m_0}$ are linearly independent. We may as well assume that the basis $E_1, \ldots, E_{m_0}$ of $\mathcal{M}_0$ is ordered so that the first p of these forms are linearly independent on S, and then we may take $\{\phi^\alpha = u^*\theta^\alpha, \alpha = 1, \ldots, p\}$ for the co-frame field on S.

Now suppose that $L_1 = \lambda_0^{-1}W_1$ is the set of first order frames on f. Then $\lambda_0 \circ u$ does not depend on the choice of <u>first</u> order frame u, which means, (using the notation of (2) in Lemma 1), that $[X] = \lambda_0 \circ u$ is independent of the choice of u. But now $x_\beta^\alpha = \delta_\beta^\alpha$ for $1 \leqslant \alpha,\beta \leqslant p$ and so $X = \begin{pmatrix} I_p \\ X_0 \end{pmatrix}$ where $I_p = p \times p$ identity matrix and $X_0 = (x_\alpha^i)_{\substack{1\leqslant\alpha\leqslant p \\ p+1\leqslant i\leqslant m_0}} : S \to \mathbb{R}^{(m_0-p)\times p}$. Thus X_0 is just the local representation of $\lambda_0 \circ u$ with respect to the standard coordinate system on $G_{m_0,p}$.

By the implicit function theorem we may choose as a coordinate system on W_1 the restriction to W_1 of an appropriate choice of $\mu_1 = \dim W_1$ of the coordinate functions of the standard coordinate system. Then on W_1, all the standard coordinate functions can be expressed as smooth functions of these coordinates of W_1. Explicitly, if $\{x^1, \ldots, x^{\mu_1}\} \subseteq \{t_\alpha^i\colon 1 \leqslant \alpha \leqslant p,\ p+1 \subseteq i \subseteq m_0\}$ form a coordinate system on W_1, then $t_\alpha^i|_{W_1} = F_\alpha^i(x^1, \ldots, x^{\mu_1})$ for certain smooth functions F_α^i of μ_1 real variables.

Lemma 2. Let α range in $\{1, \ldots, p\}$ and i range in $\{p+1, \ldots, m_0\}$. Then the first order invariants of f are a subset $\{k^1, \ldots, k^{m_1}\}$ of $\{x^i_\alpha\}$, and $x^i_\alpha = F^i_\alpha(k^1, \ldots, k^{\mu_1})$, where the F^i_α are the smooth functions determined by W_1.

The Maurer-Cartan equations of G impose relations among the invariants as follows: since $x^i_\alpha = F^i_\alpha(k^1, \ldots, k^{\mu_1})$, we have $u^*\theta^i = \sum_\alpha F^i_\alpha(k^1, \ldots, k^{\mu_1})\phi^\alpha$. Differentiating both sides, and using the Maurer-Cartan equations for $d\theta^i$, gives rise to equations relating the first order invariants and their covariant derivatives with the coefficients of $u^*\omega_0$, where ω_0 is the $\mathcal{g}_0$-component of Ω. When applying the frame construction to a specific example, one always computes these relations at each step.

We shall derive an expression for $\lambda_1 : L_1 \to G_{m_1+\mu_1,p}$ in terms of coefficients of Ω. We use the notation of section I.7.

For each $x \in W_1$, the decomposition $\mathfrak{g}_0 = \mathfrak{g}_1(x) + \mathfrak{m}_1(x)$ decomposes ω_0 into $\omega_1(x) + \Theta_1(x)$, where $\omega_1(x)$ is the $\mathfrak{g}_1(x)$-component and $\Theta_1(x)$ is the $\mathfrak{m}_1(x)$-component of ω_0. If $E_{m_0+1}(x), \ldots, E_{m_1}(x)$ is the chosen basis of $\mathfrak{m}_1(x)$, then $\Theta_1(x) = \sum_{i=m_0+1}^{m_1} \theta^i(x) \otimes E_i(x)$, where $\theta^i(x)$, $i = m_0+1, \ldots, m_1$ are ordinary left-invariant 1-forms on G, varying smoothly with $x \in W_1$; i.e., $\theta^i : W_1 \to \mathfrak{g}^*$ = dual space of $\mathfrak{g}$ is smooth.

Let u be a first order frame field along f. Let $X_0: S \to \mathbb{R}^{(m_0-p)\times p}$ be defined as in the paragraph preceding Lemma 2. For $x \in W_1$ and $m_0 + 1 \leq i \leq m_1$, set $u^*\theta^i(x) = \sum_{\alpha=1}^{p} x^i_\alpha \phi^\alpha$, where the x^i_α are smooth functions on $S \times W_1$ and $\{\phi^\alpha = u^*\theta^\alpha : \alpha = 1, \ldots, p\}$ is the coframe field on S. Let $X_1 = (x^i_\alpha)^{m_0+1\leq i\leq m_1}_{1\leq\alpha\leq p} : S \times W_1 \to \mathbb{R}^{(m_1-m_0)\times p}$

Let the subset $\{k^1, \ldots, k^{\mu_1}\}$ denote the first order invariants of f, and set $dk^a = \sum_{\alpha=1}^{p} k^a_{;\alpha}\phi^\alpha$, $1 \in a \in \mu_1$ where the $k^a_{;\alpha}$ are smooth functions on S, which we call the covariant derivatives of k^a with respect to the coframe field $\{\phi^\alpha\}$. Let $Y_1 = (k^a_{;\alpha}): S \to \mathbb{R}^{\mu_1\times p}$.

Lemma 3. If u is a first order frame field along f and X_0, X_1, and Y_1 are defined as above, then $\lambda_1 \circ u: S \to G_{m_1+\mu_1,p}$ is given by

$$\lambda_1 \circ u(s) = \begin{bmatrix} I_p \\ X_0(s) \\ X_1(s, \lambda_0 \circ u(s)) \\ Y_1(s) \end{bmatrix}, \quad s \in S .$$

Proof. Recall that $\lambda_1 \circ u(s) = (h_1 \circ u(s))^{-1}(T_f)_* S_s$, and $T_f: S \to G/G_1 \times W_1$ (using the abuse of notation in I.7) is given by $T_f(s) = (\pi_1(\lambda_0(u(s)))u(s), \lambda_0(u(s)))$, $s \in S$, where u is any first order frame field along f, and for each $x \in W_1$, $\pi_1(x): G \to G/G_1(x)$ is the natural projection.

But, as in Lemma 1, for any $x \in W_1$, $(\pi_1(x) \circ u)_*$ $= \sum_{i=1}^{m_1} (u^*\theta^i(x)) \otimes e_i$, where $e_i = u(s)_*\pi_1(x)_*E_i(x)$, $i = 1, \ldots, m_1$, where here $u(s)_*$ denotes the differential of the left translation of G/G_1 by $u(s) \in G$. Furthermore, using the local representation $(k^1, \ldots, k^{\mu_1})$ of $\lambda_0 \circ u$ defined by the chosen coordinate system $x^1, \ldots, x^{\mu_1}$ on W_1, it follows that $(T_f)_* = (\pi_1 \circ u)_* \oplus (\lambda_0 \circ u)_*$

$$= \left(\sum_{i=1}^{m_1} (u^*\theta^i) \otimes e_i\right) \oplus \sum_{a=1}^{\mu_1} dk^a \oplus \frac{\partial}{\partial x^a}$$

$$= \left(\sum_{i=1}^{m_1} \sum_{\alpha=1}^{p} x^i_\alpha \phi^\alpha \otimes e_i\right) \oplus \left(\sum_{a=1}^{\mu_1} \sum_{\alpha=1}^{p} k^a_{;\alpha}\phi^\alpha \otimes \frac{\partial}{\partial x^a}\right).$$

The lemma now follows, since $h_1 \circ u = (e_1, \ldots, e_{m_1}, \frac{\partial}{\partial x^1}, \ldots, \frac{\partial}{\partial x^{\mu_1}})$. //

Suppose that $L_2 = \lambda_1^{-1}W_2$ is the set of second order frames on f, where $W_2 \subseteq G_{m_1+\mu_1,p}$. Choose coordinates $\{x^1, \ldots, x^{\mu_2}\}$ $\subseteq \{t^i_\alpha: 1 \leqslant \alpha \leqslant p,\ p + 1 \leqslant i \leqslant m_1 + \mu_1\}$ on W_2, such that $x^1, \ldots, x^{\mu_1}$ are the ones chosen on W_1. Then $t^i_\alpha\big|_{W_2} = F^i_\alpha(x^1, \ldots, x^{\mu_2})$, $1 \leqslant \alpha \leqslant p,\ m_0 + 1 \leqslant i \leqslant m_1 + \mu_1$, for certain smooth functions F^i_α of μ_2 real variables.

If u is any second order frame field along f, then the local representation of $\lambda_1 \circ u: S \to W_2$ with respect to the above coordinate system on W_2 is $(k^1, \ldots, k^{\mu_2})$, where $k^1, \ldots, k^{\mu_1}$ are the first order invariants of f and $k^{\mu_1+1}, \ldots, k^{\mu_2}$ are the second order invariants of f. Combining this with Lemma 3 gives:

<u>Lemma 4</u>. The second order invariants $\{k^{\mu_1+1}, \ldots, k^{\mu_2}\}$ of f are a subset of $\{x^i_\alpha(s, \lambda_1 \circ u(s)) : 1 \leq \alpha \leq p,\ m_0 + 1 \leq i \leq m_1\}$ $\cup \{k^a_{;\alpha} : 1 \leq \alpha \leq p,\ 1 \leq a \leq \mu_1\}$, and $k^a_{;\alpha} = F^{m_1+a}_\alpha(k^1, \ldots, k^{\mu_2})$ for $1 \leq \alpha \leq p$, $1 \leq a \leq \mu_1$; $x^i_\alpha = F^i_\alpha(k^1, \ldots, k^{\mu_2})$ for $1 \leq \alpha \leq p$, $m_0 + 1 \leq i \leq m_1$; where the F^i_α are the smooth functions determined by W_2.

This process is iterated. We summarize the situation at the jth stage, $j \geq 1$. The jth order frames $L_j = \lambda^{-1}_{j-1} W_j$. The decomposition $\mathcal{G}_{j-1}(x) = \mathcal{G}_{j(x)} + \mathcal{M}_j(x)$, for $x \in W_j$, decomposes ω_{j-1} into $\omega_{j-1} = \omega_j + \Theta_j$ and $\Theta_j = \sum_{i=m_{j-1}+1}^{m_j} \theta^i \otimes E_i$, where $\{E_i(x) : m_{j-1} + 1 \leq i \leq m_j\}$ is the basis of $\mathcal{M}_j(x)$. If u is a jth order frame field along f, then $u^*\theta^i = \sum_{\alpha=1}^{p} x^i_\alpha \phi^\alpha$, where x^i_α is a smooth function on $S \times W_j$, $1 \leq i \leq m_j$. If $k^1, \ldots, k^{\mu_j}$ are the invariants of orders $\leq j$, then $dk^a = \sum_1^p k^a_{;\alpha}\phi^\alpha$, for smooth functions $k^a_{;\alpha}$ on S, $1 \leq a \leq \mu_j$.

Choose for coordinates on $W_{j+1} \subseteq G_{m_j+\mu_j,p}$ a subset of the standard coordinates on $G_{m_j+\mu_j,p}$, extending the set $\{x^1, \ldots, x^{\mu_j}\}$ of coordinates on W_j. Then $t^i_\alpha|_{W_{j+1}} = F^i_\alpha(x^1, \ldots, x^{\mu_{j+1}})$ for some smooth functions F^i_α of μ_{j+1} real variables, $p + 1 \leq i \leq m_j + \mu_j$, $1 \leq \alpha \leq p$. For $i \leq m_{j-1} + \mu_{j-1}$ the functions F^i_α are the ones determined by W_j .

Proposition. The $(j+1)^{st}$ order invariants $\{k^{\mu_j+1}, \ldots, k^{\mu_{j+1}}\}$ of f are a subset of $\{x^i_\alpha(s, \lambda_j \circ u(s)) : 1 \leqslant \alpha \leqslant p\ ,\ m_{j-1}+1 \leqslant i \leqslant m_j\} \cup \{k^a_{;\alpha} : 1 \leqslant \alpha \leqslant p\ ,\ \mu_{j-1} < a \leqslant \mu_j\}$, and $x^i_\alpha = F^{i+\mu_{j-1}}_\alpha(k^1, \ldots, k^{\mu_{j+1}})$ for $1 \leqslant \alpha \leqslant p\ ,\ m_{j-1}+1 \leqslant i \leqslant m_j$; $k^a_{;\alpha} = F^{m_j+a}(k^1, \ldots, k^{\mu_{k+1}})$ for $1 \leqslant \alpha \leqslant p\ ,\ \mu_{j-1}+1 \leqslant a \leqslant \mu_j$; where the F^i_α are the smooth functions determined by W_{j+1} .

I.11. Congruence and Existence Theorems. The following Theorems 1 and 2 provide the foundation for Cartan's congruence and existence theorems for submanifolds of homogeneous spaces. (Cf. Cartan [C, 1937] pp. 88-90 and p.191; Griffiths [G, 1974], pp. 780-782; or Spivak [S, 1970], Vol. I, pp. 10.40-10.44).

Theorem 1. Let $u, \tilde{u} : S \to G$ be two smooth maps of a connected manifold S into a Lie group G . Then $u(s) = A\,\tilde{u}(s)$ for fixed $A \in G$ iff $u^*\Omega = \tilde{u}^*\Omega$ where Ω is the Maurer-Cartan form of G .

Theorem 2. Suppose that Φ is a $\mathfrak{g}$-valued 1-form on a domain S in $\mathbb{R}^p$, where $\mathfrak{g}$ is the Lie algebra of G . Then for any point $s_0 \in S$ there exists a neighborhood S' of s_0 in S and a smooth map $u: S' \to G$ such that $u^*\Omega = \Phi$ iff $d\Phi = -\frac{1}{2}[\Phi, \Phi]$.

<u>Corollary 1</u>. Let $f: S \to M$ and $\tilde{f}: \tilde{S} \to M$ be smoothly imbedded p-dimensional submanifolds of the homogeneous $M = G/G_0$. Then f and $\tilde{f}$ are G-congruent iff there exist zeroth order frame fields $u: S \to G$ and $\tilde{u}: \tilde{S} \to G$ along f and $\tilde{f}$, respectively, and a diffeomorphism $F: S \to \tilde{S}$ such that $u^*\Omega = (\tilde{u} \circ F)^*\Omega$.

<u>Proof</u>: Suppose f and $\tilde{f}$ are G-congruent. Then there exists $A \in G$ such that $A\, f(S) = \tilde{f}(\tilde{S})$. This defines a diffeomorphism $F: S \to \tilde{S}$ such that $A \circ f = \tilde{f} \circ F: S \to M$. Let $\tilde{u}$ be a zeroth order frame field along $\tilde{f}$. Then $\tilde{u} \circ F$ is a zeroth order frame field along $\tilde{f} \circ F = A \circ f$, so that $u \equiv A^{-1} \cdot (\tilde{u} \circ F)$ is a zeroth order frame field along f. Then, by the left-invariance of Ω, $u^*\Omega = (\tilde{u} \circ F)^*\Omega$.

Conversely, suppose u, $\tilde{u}$ and F exist, as stated, and that $u^*\Omega = (\tilde{u} \circ F)^*\Omega$. By Theorem 1, there is an $A \in G$ such that $u(s) = A \cdot (\tilde{u} \circ F)(s)$, for all $s \in S$. But then $f(s) = \pi\, u(s) = \pi(A \cdot \tilde{u}(F(s))) = A \cdot \pi(\tilde{u}(F(s)))) = A \cdot \tilde{f}(F(s))$, for all $s \in S$, and thus $f(S) = A\tilde{f}(\tilde{S})$. //

Let $f: S \to M$ and $\tilde{f}: \tilde{S} \to M$ be p-dimensional smoothly imbedded submanifolds. Let L_j denote the jth order frames on f with respect to the local cross section $W_j \subseteq G_{m_j+\mu_j,p}$. Let $k^1, \ldots, k^{\mu_j}$ denote the invariants of orders 1 through j of f, $j = 1, 2, 3, \ldots$. Denote these quantities for $\tilde{f}$ by the same letters with a tilde.

The following theorem is trivial for real analytic submanifolds. We shall give a proof which is valid when f and $\tilde{f}$ are smooth, i.e. of class C^{q+1} in this case.

<u>Theorem 3</u>. (Cartan [C, 1937], p. 154) f and $\tilde{f}$ are G-congruent iff (i) their Frenet frames are of order q, (ii) they are both the type of W_q, and (iii) there exists a one-to-one correspondence $\psi\colon S \to \tilde{S}$ such that $\tilde{k}^a \circ \psi = k^a$, $0 \leqslant a \leqslant \mu_{q+1}$. (To be precise, we mean by the converse that (i), (ii) and (iii) imply that for any point $s \in S$ there are neighborhoods S' of 0 and $\tilde{S}'$ of $\psi(s)$, and there is an element $A \in G$, such that $A\, f(S') = \tilde{f}(\tilde{S}')$.)

<u>Proof</u>. The necessity of (i), (ii) and (iii) for G-congruence is obvious. To prove sufficiency, suppose now that (i), (ii) and (iii) hold.

Choose and fix Frenet frame fields u and $\tilde{u}$ along f and $\tilde{f}$, respectively. In order to prove the theorem it suffices, by Corollary 1, to find a diffeomorphism $F\colon S \to \tilde{S}$ and a smooth map $K\colon S \to G_q$, (i.e., $K\colon S \to G$ such that $K(s) \in G_q(\lambda_q \circ u(s))$, for each $s \in S$), such that $F^*\tilde{u}^*\Omega = (uK)^*\Omega$. We begin by developing some notation.

For the remainder of this proof, except where explicitly countermanded, we adopt the following index convention: $1 \leqslant a, b, c, d, e \leqslant r = \dim G$; $1 \leqslant i, j, k \leqslant m_q$; $m_{q+1} \leqslant \lambda, \mu, \nu, \tau \leqslant r$; $1 \leqslant \alpha, \beta, \gamma, \delta, \varepsilon \leqslant p$; $1 \leqslant A, B \leqslant \mu_q$.

By definition of the Frenet frame, all invariants of f are functions of the invariants of orders $\leqslant q$, namely $\{k^A\}$. Let $n \in \{0, 1, \ldots, p\}$ be the maximal number of independent invariants on S, and let $\{s^1, \ldots, s^n\} \subseteq \{k^A\}$ denote a set of independent invariants. Then $k^A = h^A(s^1, \ldots, s^n)$ for some smooth functions h^A of n real variables. By hypothesis (iii) any functional relationship among the $\{k^A\}$ also exists among the $\{\tilde{k}^A\}$, and conversely. Thus

$\{\tilde{s}^1, \ldots, \tilde{s}^n\}$, where $\tilde{s}^i = s^i \circ \psi$, $i = 1, \ldots, n$, is a maximal set of independent invariants on $\tilde{S}$, and $\tilde{k}^A = h^A(\tilde{s}^1, \ldots, \tilde{s}^n)$, for the same functions h^A .

The frame construction has given rise to a decomposition $\mathfrak{g} = \mathfrak{g}_q \oplus \mathfrak{m}$, where $\mathfrak{m} = \mathfrak{m}_0 \oplus \ldots \oplus \mathfrak{m}_q$. (The subspaces $\mathfrak{m}_j$ and $\mathfrak{g}_j$ may depend smoothly on points in W_q .) Combining the bases of each $\mathfrak{m}_j$, we have a basis $E_1, \ldots, E_{m_q}$ of $\mathfrak{m}$, which we extend to a basis of $\mathfrak{g}$ by choosing a basis $E_{m_{q+1}}, \ldots, E_r$ of $\mathfrak{g}_q$.

The Maurer-Cartan form Ω of G decomposes into $\Omega = \omega_q + \Theta$, where ω_q is the $\mathfrak{g}_q$-component and Θ is the $\mathfrak{m}$-component. Set $\Theta = \sum \theta^i \otimes E_i$ and $\omega_q = \sum \theta^\lambda \otimes E_\lambda$, so $\Omega = \sum \theta^a \otimes E_a$, where the θ^a are left-invariant 1-forms on G , (depending smoothly on $x \in W_q$) . The Maurer-Cartan form of G_q may be expressed by $\sum \kappa^\lambda \otimes E_\lambda$, where the κ^ν are left-invariant 1-forms on G_q .

Let $u^*\theta^i = \sum x^i_\alpha \phi^\alpha$, and $dk^A = \sum k^A_{;\alpha} \phi^\alpha$, where $\{\phi^\alpha = u^*\theta^\alpha\}$ is a coframe field on S . Then each $x^i_\alpha = F^i_\alpha(k^1, \ldots, k^{\mu_q})$ and $k^A_{;\alpha} = N^A_\alpha(k^1, \ldots, k^{\mu_q})$, where the smooth functions F^i_α and N^A_α depend only on W_q , <u>not</u> on f (as emphasized in the frame construction, cf. the Proposition in I.10).

Denote all the above quantities of $\tilde{u}$ on $\tilde{f}$ by the same letters with tildes. Then $\tilde{x}^i_\alpha = F^i_\alpha(\tilde{k}^1, \ldots, \tilde{k}^{\mu_q})$ and $\tilde{k}^A_{;\alpha} = N^A_\alpha(\tilde{k}^1, \ldots, \tilde{k}^{\mu_q})$, so $\tilde{x}^i_\alpha \circ \psi = x^i_\alpha$ and $\tilde{k}^A_{;\alpha} \psi = k^A_{;\alpha}$. Thus whenever $s \in S$, $\tilde{s} \in \tilde{S}$ are such that $s^i(s) = \tilde{s}^i(\tilde{s})$ for $i = 1, \ldots, n$, then $x^i_\alpha(s) = \tilde{x}^i_\alpha(\tilde{s})$ and $k^A_{;\alpha}(s) = \tilde{k}^A_{;\alpha}(\tilde{s})$.

The structure constants C^a_{bc} of $\mathfrak{g}$ are defined by $[E_b, E_c] = \sum C^a_{bc} E_a$, or equivalently $d\theta^a = -\frac{1}{2}\sum C^a_{bc}\theta^b \wedge \theta^c$. For any $K \in G_q$ we set $Ad(K^{-1})E_a = \sum H^b_a E_b$. If K depends smoothly on $s \in S$, then the coefficients H^b_a are smooth functions on S.

Since the qth order frames are the Frenet frames, the action ρ_q of G_q on $W_q \subseteq G_{m_q+\mu_q,p}$ must be trivial. Consequently, for each i, $\sum H^i_j x^j_\alpha = \sum A^\beta_\alpha x^i_\beta$, where the non-singular $p \times p$ matrix $A = (A^\alpha_\beta)$ is determined by $K \in G_q$. Then A is just the change of basis matrix from $\phi^1, \ldots, \phi^p$ to $\hat{\phi}^1, \ldots, \hat{\phi}^p$, where $\hat{\phi}^\alpha = (uK)^*\theta^\alpha$, i.e., $\hat{\phi}^\alpha = \sum A^\alpha_\beta \phi^\beta$. In fact, $(uK)^*\Theta = (Ad(K^{-1})\sum u^*\theta^i \otimes E_i)_{\mathfrak{m}} = \sum u^*\theta^i \otimes H^j_i E_j$ implies $\hat{\phi}^\alpha = (uK)^*\theta^\alpha = \sum H^\alpha_i u^*\theta^i = \sum H^\alpha_i x^i_\beta \phi^\beta = \sum x^\alpha_\gamma A^\gamma_\beta \phi^\beta = \sum A^\alpha_\beta \phi^\beta$, since $x^\alpha_\gamma = \delta^\alpha_\gamma$.

Suppose for the moment that smooth maps $F: S \to \tilde{S}$ and $K: S \to G_q$ exist such that $F^*\tilde{u}^*\Omega = (uK)^*\Omega$. Then

(1) $F^*\tilde{u}^*\Theta = (Ad(K^{-1})u^*\Theta)_{\mathfrak{m}}$, and

(2) $F^*\tilde{u}^*\omega_q = (Ad(K^{-1})u^*\Omega)_{\mathfrak{g}_q} + K^*(\sum \kappa^\lambda \otimes E_\lambda)$.

To express (1) in terms of the basis, observe above that $(Ad(K^{-1})u^*\Theta)_{\mathfrak{m}} = \sum H^j_i x^i_\alpha \phi^\alpha \otimes E_j = \sum x^j_\beta A^\beta_\alpha \phi^\alpha \otimes E_j$, while $F^*\tilde{u}^*\Theta = \sum F^*(\tilde{x}^j_\beta \tilde{\phi}^\beta) \otimes E_j$. Since F must preserve the invariants, i.e. $\tilde{k}^A \circ F = k^A$, it follows $F^*\tilde{x}^j_\beta = x^j_\beta$, and hence (1) is implied by

(3) $F^*\tilde{\phi}^\beta = \sum A^\beta_\alpha \phi^\alpha$.

Turning to (2), $(Ad(K^{-1})u^*\Omega)_{\mathfrak{g}_q} + K^*(\sum \kappa^\lambda \otimes E_\lambda) = \sum (H^\alpha_a u^*\theta^a + K^*\kappa^\lambda) \otimes E_\lambda$, while $F^*\tilde{u}^*\omega_q = \sum F^*\tilde{u}^*\theta^\lambda \otimes E_\lambda$. Hence (2) is implied by

(4) $F^*\tilde{u}^*\theta^\lambda = \sum H^\lambda_a u^*\theta^a + K^*\kappa^\lambda$.

Thus F and K are solutions of the system of differential equations (3) and (4). We show solutions to (3) and (4) exist by the standard technique of constructing the graph, using an appropriately defined distribution on $S \times \tilde{S} \times G_q$.

Let X be the $2p - n$ dimensional submanifold of $S \times \tilde{S}$ defined by the equations $\{s^i = \tilde{s}^i : i = 1, \ldots, n\}$. Notice that the projection map $S \times \tilde{S} \to S$ when restricted to X continues to have rank p at every point of X . By an abuse of notation we let $X \times G_q = \bigcup_{(s,\tilde{s}) \in X} \{(s, \tilde{s})\} \times G_q(\lambda_q \circ u(s)) \subseteq X \times G$. (There is no abuse if G_q does not depend on $x \in W_q$.)

On $X \times G_q$ define the 1-forms:

(5) $\eta^\alpha = u^*\theta^\alpha - (uK)^*\theta^\alpha$, i.e., $\eta^\alpha = \tilde{\phi}^\alpha - \sum A^\alpha_\beta \phi^\beta$; and

(6) $\psi^\lambda = \tilde{u}^*\theta^\lambda - (uK)^*\theta^\lambda$, i.e., $\psi^\lambda = \tilde{u}^*\theta^\lambda - \sum H^\lambda_a u^*\theta^a - \kappa^\lambda$.

Let D denote the smooth distribution on $X \times G_q$ defined by the equations $\{\eta^\alpha = 0 , \ \psi^\lambda = 0 : \ \alpha = 1, \ldots, p ; \ \lambda = m_q + 1, \ldots, r\}$. To complete the proof it suffices to show that D is a p-dimensional involutive distribution such that $D(s, \tilde{s}, K)$ projects onto the tangent space S_s for each point $(s, \tilde{s}, K) \in X \times G_q$. We do this in the following three lemmas.

<u>Lemma A</u>. Reordering $\theta^1, \ldots, \theta^p$, if necessary, we may assume $ds , \ldots, ds^n , \phi^{n+1}, \ldots, \phi^p$ is a coframe field on S , i.e. linearly independent at every point. Then $d\tilde{s}^1, \ldots, d\tilde{s}^n , \tilde{\phi}^{n+1}, \ldots, \tilde{\phi}^p$ is a coframe field on $\tilde{S}$, and $ds^1, \ldots, ds^n , \hat{\phi}^1, \ldots, \hat{\phi}^p$ is

a coframe field on S , where $\hat{\phi}^{\alpha} = (uK)^{*}\phi^{\alpha}$ for any $K: S \to G_q$.

<u>Proof</u>. Let the index t range from $1, \ldots, n$. Then $ds^t = \sum s^t_{;\alpha}\phi^{\alpha}$, where the functions $s^t_{;\alpha} \in \{k^A_{;\beta}\}$. Let T denote the $n \times \rho$ matrix of functions $(s^t_{;\alpha})$. Then $ds^1, \ldots, ds^n, \phi^{n+1}, \ldots, \phi^p$ are linearly independent iff the first n columns of T are linearly independent. Since the $s^t_{;\alpha}$ are invariants, we have $ds^t = \sum s^t_{;\alpha}\hat{\phi}^{\alpha}$, where $\hat{\phi}^{\alpha} = \hat{u}^{*}\theta^{\alpha}$ for any Frenet frame field $\hat{u}$ along f . Hence $ds^1, \ldots, ds^n$, $\hat{\phi}^{n+1}, \ldots, \hat{\phi}^p$ is a coframe field on S .

Finally, $d\tilde{s}^t = \sum \tilde{s}^t_{;\alpha}\tilde{\phi}^{\alpha}$, and $\tilde{T} = (\tilde{s}^t_{;\alpha})$, then $\tilde{s}^t_{;\alpha} \circ \psi = s^t_{;\alpha}$ and $\tilde{T} \circ \psi = T$. Thus the first n columns of $\tilde{T}$ are linearly independent, and $d\tilde{s}^1, \ldots, d\tilde{s}^n$, $\tilde{\phi}^{n+1}, \ldots, \tilde{\phi}^p$ is a coframe field on S .//

<u>Lemma B</u>. D is a smooth p-dimensional distribution on $X \times G_q$ such that $D(s, \tilde{s}, K)$ projects onto the tangent space S_s , for each $(s, \tilde{s}, K) \in X \times G_q$.

<u>Proof</u>. To compute the dimension of D , observe that the $\{\psi^{\lambda}: \lambda = m_q + 1, \ldots, r\}$ are linearly independent, since the κ^{λ} are; and $\eta^{n+1}, \ldots, \eta^p$ are linearly independent while $\eta^1, \ldots, \eta^n$ are linear combinations (with function coefficients) of $\eta^{n+1}, \ldots, \eta^p$. To see this, we continue using the notaiton in the proof of Lemma A, and set $T = (a, b)$, where a is an $n \times n$ non-singular matrix, and b is an $n \times (p - n)$ matrix. The entries of a and b are the invariants $s^t_{;\alpha}$. If $\hat{u}$ is any Frenet frame field along f , and

$\hat{\phi}^\alpha = \hat{u}^*\theta^\alpha$, then $\begin{pmatrix}\hat{\phi}^1\\ \vdots \\ \hat{\phi}^n\end{pmatrix} = a^{-1}\begin{pmatrix}ds^1\\ \vdots \\ ds^n\end{pmatrix} - a^{-1}b\begin{pmatrix}\hat{\phi}^{n+1}\\ \vdots \\ \hat{\phi}^p\end{pmatrix}$. Similarly, if $\tilde{a} = a \circ \psi^{-1}$ and $\tilde{b} = b \circ \psi^{-1}$, so $\tilde{T} = (\tilde{a}, \tilde{b})$, then $\begin{pmatrix}\tilde{\phi}^1\\ \vdots \\ \tilde{\phi}^n\end{pmatrix} = \tilde{a}^{-1}\begin{pmatrix}d\tilde{s}^1\\ \vdots \\ d\tilde{s}^n\end{pmatrix} - \tilde{a}^{-1}\tilde{b}\begin{pmatrix}\tilde{\phi}^{n+1}\\ \vdots \\ \tilde{\phi}^p\end{pmatrix}$.

But on X , $\tilde{T} = T$, $\tilde{s}^i = s^i$, and $\begin{pmatrix}\eta^1\\ \vdots \\ \eta^n\end{pmatrix} = \begin{pmatrix}\tilde{\phi}^1\\ \vdots \\ \tilde{\phi}^n\end{pmatrix} - \begin{pmatrix}\hat{\phi}^1\\ \vdots \\ \hat{\phi}^n\end{pmatrix} = -a^{-1}b\begin{pmatrix}\eta^{n+1}\\ \vdots \\ \eta^p\end{pmatrix}$. Finally, $\eta^{n+1}, \ldots, \eta^p$ are linearly independent by Lemma A. Since D is defined by $r - m_q + p - n$ independent equations, its dimension is $\dim X - (r - m_q + p - n) = 2p - n + r - m_q - (r - m_q + p - n) = 0$.

Take $(s, \tilde{s}, K) \in X \times G_q$ and take $v \in S_s$. Let $e_1, \ldots, e_p$ be the basis of S_s dual to $ds^1, \ldots, ds^n$, $\hat{\phi}^{n+1}, \ldots, \hat{\phi}^p$, where $\hat{\phi}^\alpha = \sum A^\alpha_\beta \phi^\beta$. Let $\tilde{e}_1, \ldots, \tilde{e}_p$ be the basis of $\tilde{S}_{\tilde{s}}$ dual to $d\tilde{s}^1, \ldots, d\tilde{s}^n$, $\tilde{\phi}^{n+1}, \ldots, \tilde{\phi}^p$. Then $v = \sum v^\alpha e_\alpha$, for some $v^\alpha \in \mathbb{R}$. If we let $\tilde{v} = \sum v^\alpha \tilde{e}_\alpha \in \tilde{S}_{\tilde{s}}$ and let $w = \sum v^\lambda E_\lambda \in \mathcal{F}_q$, where $v^\lambda = \tilde{u}^*\theta^\lambda(\tilde{v}) - \sum H^\lambda_a u^*\theta^a(v) \in \mathbb{R}$, then clearly $(v, \tilde{v}, w) \in D(s, \tilde{s}, K)$ and this vector projects onto v . This completes the proof of Lemma B.//

<u>Lemma C.</u> D is involutive.

<u>Proof.</u> It suffices to show $d\eta^\alpha$ and $d\psi^\lambda$ are zero modulo $(\eta^1, \ldots, \eta^p , \psi^{m_q+1}, \ldots, \psi^r)$.

Now $\eta^\alpha = \tilde{\phi}^\alpha - \sum A^\alpha_\beta \phi^\beta = \tilde{\phi}^\alpha - \sum H^\alpha_i x^i_\beta \phi^\beta = \tilde{\phi}^\alpha - \sum H^\alpha_i u^*\theta^i$.

(7) $d\tilde{\phi}^\alpha = -\frac{1}{2} \sum C^\alpha_{ij} \tilde{x}^i_\beta x^j_\gamma \tilde{\phi}^\beta \wedge \tilde{\phi}^\gamma - \sum C^\alpha_{i\lambda} \tilde{x}^i_\beta \tilde{\phi}^\beta \wedge \tilde{u}^*\theta^\lambda$, (since $C^\alpha_{\lambda\mu} = 0$) ;

(8) $dH^a_b = -\sum C^a_{\nu c} H^c_b \kappa^\nu$, since $\mathrm{Ad}(K^* \sum \kappa^\lambda \otimes E_\lambda) = \mathrm{Ad}(K^{-1}) d\mathrm{Ad}(K)$;

(9) $du^*\theta^i = -\frac{1}{2} \sum C^i_{jk} x^j_\alpha x^k_\beta \phi^\alpha \wedge \phi^\beta - \sum C^i_{j\lambda} x^j_\alpha \phi^\alpha \wedge u^*\theta^\lambda$.

(10) On X , $\tilde{x}^i_\beta = x^i_\beta$, and $\tilde{\phi}^\alpha = \sum H^\alpha_i u^*\theta^i$, $\kappa^\lambda = \tilde{u}^*\theta^\lambda - \sum H^\lambda_a u^*\theta^a$

modulo $\{\eta^\alpha, \psi^\lambda\}$. Hence

$$\begin{aligned} d\eta^\alpha = &- \frac{1}{2} \sum C^\alpha_{ij} x^i_\beta x^j_\gamma \tilde{\phi}^\beta \wedge \tilde{\phi}^\gamma - \sum C^\alpha_{i\lambda} x^i_\beta \tilde{\phi}^\beta \wedge \tilde{u}^* \theta^\lambda + \sum H^i C^\alpha_{\nu i} x^j_\delta \kappa^\nu \wedge \phi^\delta \\ &+ \frac{1}{2} \sum H^\alpha_j C^j_{ik} x^i_\delta x^k_\varepsilon \phi^\delta \wedge \phi^\varepsilon + \sum H^\alpha_j C^j_{i\lambda} x^i_\delta \phi^\delta \wedge u^* \theta^\lambda \\ = &\{ - \frac{1}{2} \sum C^\alpha_{ij} x^i_\beta x^j_\gamma H^\beta_k H^\gamma_\ell x^k_\delta x^\ell_\varepsilon + \sum H^i_j C^\alpha_{\nu i} x^j_\delta H^\nu_k x^k_\varepsilon + \frac{1}{2} \sum H^\alpha_j C^j_{ik} x^i_\delta x^k_\varepsilon \} \phi^\delta \wedge \phi^\varepsilon \\ &+ \{ - \sum C^\alpha_{i\lambda} H^\beta_j x^i_\beta x^j_\delta - \sum C^\alpha_{\lambda i} H^i_j x^j_\delta \} \phi^\delta \wedge \tilde{u}^* \theta^\lambda \\ &+ \{ - \sum C^\alpha_{i\nu} H^i_j H^\nu_\lambda x^j_\delta + \sum C^j_{i\lambda} H^\alpha_j x^i_\delta \} \phi^\delta \wedge u^* \theta^\lambda \bmod \{\eta^\alpha, \psi^\lambda\} . \end{aligned}$$

Since $Ad(K^{-1})$ is an automorphism on $\mathfrak{g}$, we have the identity

$$(11) \quad \sum H^a_b H^d_c C_{ad} = \sum H^e_a C^a_{bc} .$$

Furthermore, since $\mathfrak{g}_q$ is a Lie subalgebra and $K^{-1} \in G_q$, we have

$$(12) \quad C^i_{\lambda\mu} = 0 \quad \text{and} \quad H^i_\lambda = 0 .$$

Coefficient of $\phi^\delta \wedge u^* \theta^\lambda$ is $- \sum C^\alpha_{ab} H^a_j H^b_\lambda x^j_\delta + \sum C^j_{i\lambda} H^\alpha_j x^i_\delta$
$= - \sum H^\alpha_a C^a_{j\lambda} x^j_\delta + \sum C^j_{i\lambda} H^\alpha_j x^i_\delta = 0$, by (11).
Coefficient of $\phi^\delta \wedge \tilde{u}^* \theta^\lambda$ is $- \sum C^\alpha_{i\lambda} x^i_\beta x^\beta_\gamma A^\gamma_\delta - \sum C^\alpha_{\lambda i} x^i_\gamma A^\gamma_\delta = 0$, since
$x^\beta_\gamma = \delta^\beta_\gamma$.
Coefficient of $\phi^\delta \wedge \phi^\varepsilon$ is $- \frac{1}{2} \sum C^\alpha_{ij} H^i_k H^j_\ell x^k_\delta x^\ell_\varepsilon - \sum C^\alpha_{i\nu} H^\nu_j H^i_k x^j_\varepsilon x^k_\delta + \frac{1}{2} \sum C^j_{ik} H^\alpha_j x^i_\delta x^k_\varepsilon$
$= - \frac{1}{2} \sum C^\alpha_{ab} H^a_k H^b_\ell x^k_\delta x^\ell_\varepsilon + \frac{1}{2} \sum C^j_{ik} H^\alpha_j x^i_\delta x^k_\varepsilon = 0$, by (11).
Hence $d\eta^\alpha = 0 \bmod(\eta^\alpha, \psi^\lambda)$.

Using (7), (8), (9), (10), and the Maurer-Cartan equation of G_q: $d\kappa^\lambda = -\frac{1}{2}\sum C^\mu_{\lambda\nu}\kappa^\lambda \wedge \kappa^\nu$; we get

$$d\psi^\lambda = \sum \{\tfrac{1}{2} C^\mu_{ij} H^i_k H^j_\ell x^k_\alpha x^\ell_\beta - \tfrac{1}{2} C^\mu_{k\ell} x^k_\gamma x^\ell_\delta H^\gamma_i H^\delta_j x^i_\alpha x^j_\beta\}\phi^\alpha \wedge \phi^\beta$$
$$+ \sum C^\mu_{\nu_j} \tilde{u}^*\theta^\nu \wedge (H^j_i u^*\theta^i - \tilde{u}^*\theta^j) \quad \text{modulo } \{\eta^\alpha, \psi^\lambda\} .$$

Coefficient of $\phi^\alpha \wedge \phi^\beta$ is $\frac{1}{2}\sum C^\mu_{ij} x^i_\gamma A^\gamma_\alpha x^j_\delta A^\delta_\beta - \frac{1}{2}\sum C^\mu_{k\ell} x^k_\gamma x^\ell_\delta A^\gamma_\alpha A^\delta_\beta = 0$; $\sum (H^j_i u^*\theta^i - \tilde{u}^*\theta^j) = \sum (H^j_i x^i_\alpha \phi^\alpha - x^j_\alpha \tilde{\phi}^\alpha) = \sum (x^j_\beta A^\beta_\alpha \phi^\alpha - x^j_\alpha \tilde{\phi}^\alpha) = 0$, by (10). Hence $d\psi^\lambda = 0 \mod (\eta^\alpha, \psi^\lambda)$.

This completes the proof of Lemma C, which in turn completes the proof of Theorem 3.//

<u>Remark</u>. If $f: S \to M$ and $\tilde{f}: \tilde{S} \to M$ are G-congruent, then the transformation $A \in G$ sending $f(S)$ onto $\tilde{f}(\tilde{S})$ is given by their Frenet frames. In fact, suppose $A \in G$ and $Af(S) = \tilde{f}(\tilde{S})$. If $B \in L_q$ is a Frenet frame of f at $s \in S$, then $\tilde{B} = AB \in \tilde{L}_q$ is a Frenet frame of $\tilde{f}$ at $\tilde{s} \in \tilde{S}$, where $\tilde{f}(\tilde{s}) = Af(s)$. Hence $A = \tilde{B}B^{-1}$, where B and $\tilde{B}$ are Frenet frames of f and $\tilde{f}$ at points $s \in S$ and $\psi(s) \in \tilde{S}$, respectively.

We shall continue using the notation developed above. Let $f: S \to M$ be a p-dimensional smoothly imbedded submanifold of the type of $W_q \subseteq G_{m_q+\mu_q,p}$ and with set of invariants $\{k^1, \ldots, k^{\mu_q}\}$, where q is the order of the Frenet frames of f . Let u be a Frenet frame field along f and set $u^*\theta^a = \sum x^a_\alpha \phi^\alpha$, where $\{\phi^\alpha = u^*\theta^\alpha\}$ is a coframe field on S , and $\Omega = \sum \theta^a \otimes E_a$, (see the proof of Theorem 3 for notation and index conventions). We know that $x^i_\alpha = F^i_\alpha(k^1, \ldots, k^{\mu_q})$, where the functions F^i_α are determined by W_q .

Taking the exterior derivative of both sides of $u^*\theta^a = \sum x^a_\alpha \phi^\alpha$, and using the Maurer-Cartan equations of G we get

$$(13)\quad d\phi^\alpha = -\frac{1}{2}\sum c^\alpha_{bc} x^b_\beta x^c_\gamma \phi^\beta \wedge \phi^\gamma \qquad 1 \leqslant \alpha, \beta, \gamma \leqslant p$$

$$x^a_{\alpha;\beta} - x^a_{\beta;\alpha} = -\sum x^a_\gamma c^\gamma_{bc} x^b_\alpha x^c_\beta \qquad \text{for } a = p+1, \ldots, r .$$

We call the r equations in (13) the <u>structure equations</u> of f . These are the only general relations that must hold among the invariants of f in the following sense.

<u>Theorem 4</u>. Let S be a domain containing 0 in $\mathbb{R}^p$. Let $\phi^1, \ldots, \phi^p$ be a coframe field on S , and let $k^1, \ldots, k^{\mu_q}$ and x^λ_α , $1 \leqslant \alpha \leqslant p$, $m_q + 1 \leqslant \lambda \leqslant r$, be smooth functions on S . Let $x^i_\alpha = F^i_\alpha(k^1, \ldots, k^{\mu_q})$, where F^i_α are the functions defined by W_q , $1 \leqslant i \leqslant m_q$. If the equations (13) are satisfied, then there exists a smooth imbedding $f\colon S' \to M$ of the type of W_q whose invariants of orders $\leqslant q$ are $k^1, \ldots, k^{\mu_q}$, where $0 \in S' \subseteq S$ is open.

<u>Proof</u>: Let $\phi^a = \sum x^a_\alpha \phi^\alpha$ (note $x^\beta_\alpha = \delta^\beta_\alpha$) , and let $\Phi = \sum \phi^a \otimes E_a$, a $\mathfrak{g}$-valued 1-form on S . By (13), $d\Phi = -\frac{1}{2}[\Phi, \Phi]$, so by Theorem 2 there exists a smooth map $u\colon S' \to G$ such that $u^*\Omega = \Phi$. Let $f = \pi \circ u\colon S' \to M$. It is clear from the frame construction on f that u is a qth order frame field on f with $\lambda_q \circ u(S') \subseteq W_q$ and that $k^1, \ldots, k^{\mu_q}$ are the invariants of all orders $\leqslant q$ .//

I.12. Homogeneity theorem. The theorem of this section deals with the question: when is there a subgroup of G leaving a given submanifold invariant? It is clear that an element of G which transforms a submanifold into itself will preserve the invariants, that is, the level manifolds of the invariants are preserved by the transformation. As the next theorem shows, these level manifolds are pieces of homogeneous submanifolds, all orbits of the same subgroup of G when f is analytic.

Theorem 5 (Cartan [C, 1937], p. 155). Let $f: S \to M$ be a p-dimensional smoothly imbedded submanifold, and let q be the order of its Frenet frames. Let $\{k^1, \ldots, k^{\mu_q}\}$ be the set of all its invariants of orders $\leq q$, and let $k: S \to \mathbb{R}^{\mu_q}$ be the smooth map $k = (k^1, \ldots, k^{\mu_q})$. For a regular value y of k, let $Y(y) = k^{-1}\{y\}$. Then there is a closed subgroup $H(y)$ of G such that $H(y) \supseteq G_{q0}$, where G_{q0} is the connected component of the identity of G_q, and $f(Y(y))$ is G-congruent to an open submanifold of the homogeneous submanifold $\pi H(y)$; i.e., there is an $A \in G$ such that $Af(Y(y)) \subseteq \pi H(y)$.

Furthermore, suppose f is of class C^ω. Choose a point $s_0 \in S$ where k is non-singular, and translate f by an element of G, if necessary, so that $f(s_0) = 0$ and $1 \in G$ is a Frenet frame of f at s_0. Then $H(y_0)$ is a group of local transformations of $f(S)$. In other words, $f(S)$ is an open submanifold of the p-dimensional submanifold $H(y_0)\cdot f(S)$ of M. Note that $H(y_0)\cdot f(S)$ is the union of the $H(y_0)$-orbits in M of a "cross-section" in S of the level spaces of $k: S \to \mathbb{R}^{\mu_q}$.

Corollary: For any smooth imbedding $f: S \to M$, $f(S)$ is an open submanifold of a homogeneous submanifold of M iff f has no non-constant invariants.

Proof. We use the notation and index conventions introduced in the proof of Theorem 3. Let u be a Frenet frame field along f, $\Omega = \sum_1^r \theta^a \otimes E_0$ the Maurer-Cartan form of G, and set $u^*\theta^i = \sum_1^p x^i_\alpha \theta^\alpha$, where $\{\phi^\alpha = u^*\theta^\alpha : \alpha = 1, \ldots, p\}$ is a coframe field on S. Then $x^i_\alpha = F^i_\alpha(k^1, \ldots, k^{\mu_q})$ are constant on $Y(y)$. Let $y \in k(S)$ be a regular value of k.

Let $\mathcal{L}(y)$ be the left-invariant distribution on G defined by the equations $\{\theta^i = \sum x^i_\alpha(y)\theta^\alpha$, $\sum k^a_{;\alpha}\theta^\alpha = 0 : i = p+1, \ldots, m_q$, $a = 1, \ldots, \mu_q\}$. If $n \in \{0, 1, \ldots, p\}$ is the maximal number of independent invariants on S, i.e., n is the rank of the $\mu_q \times p$ matrix $(k^a_{;\alpha})$ at any point of S, (or at least on an open submanifold of S), then there are $n + m_q - p$ independent equations defining $\mathcal{L}(y)$, so the $\dim \mathcal{L}(y) = r + p - n - m_q$. Furthermore $Y(y)$ is a smooth submanifold of S of codimension n. In particular if $n = 0$, then $Y(y) = S$.

Let $L_q \subseteq G$ denote the submanifold of qth order frames on f. The fibration $\pi: L_q \to S$ has fibre over s given by $u(s)G_q(\lambda_q \circ u(s))$, for each $s \in S$. But the map $k: S \to \mathbb{R}^{\mu_q}$ is just a local representation of $\lambda_q \circ u$ with respect to the chosen coordinate system of W_q, and hence L_q restricted to $Y(y)$, denoted $L_q(y)$, is a trivial principal bundle with structure group $G_q(y)$. Clearly $L_q(y)$ is an integral submanifold of $\mathfrak{h}(y)$, and thus the left-translates of $L_q(y)$ are integral submanifolds of $\mathfrak{h}(y)$ through any point of G . Since $\dim L_q(y) = \dim Y(y) + \dim G_q(y) = p - n + r - m_q = \dim \mathfrak{h}(y)$, it follows that $\mathfrak{h}(y)$ is involutive; i.e., $\mathfrak{h}(y)$ is a Lie subalgebra of $\mathfrak{g}$.

Let H(y) be the analytic subgroup of G whose Lie algebra is $\mathfrak{h}(y)$; that is, H(y) and its left translates are the maximal integral submanifolds of $\mathfrak{h}(y)$. H(y) contains the connected component of the identity of $G_q(y)$ since $\mathfrak{g}(y) \supseteq \mathfrak{g}_q(y)$. Since u(Y(y)) is an integral submanifold of $\mathfrak{h}(y)$, there is an $A \in G$ such that $Au(Y(y)) \subseteq H(y)$. Hence $Af(Y(y)) = A\pi u(Y(y)) = \pi Au(Y(y)) \subseteq \pi H(y)$.

Finally, suppose f is of class C^ω , and fix a point $s_0 \in Y(y)$. Translating f by an element of G , if necessary, we may assume $f(s_0) = 0$ and that $1 \in G$ is a Frenet frame of f at s_0 . Let $A \in H(y)$ be sufficiently close to 1 that $Af(s_0) \in f(S)$, and let $s_1 \in Y(y)$ be the point for which $f(s_1) = Af(s_0)$. Let $\tilde{f} = A \circ f\colon S \to M$. Then $\tilde{f}(s_0) = f(s_1)$ and the set of invariants of $\tilde{f}$ at s_0 is the same as the set of invariants of f at s_0 , which in turn is the same as the set of invariants of f at s_1 , since $k(s_0) = k(s_1)$. Hence f and $\tilde{f}$ have infinite order contact at s_1 and s_0 , and thus $f(S') = \tilde{f}(S'')$ for some open neighborhoods S' and S'' of s_1 and s_0 , respectively, in S , since they are real analytic maps. Then $Af(S'') = \tilde{f}(S'') = f(S') \subseteq f(S)$, which is what we wanted to prove.

II. Surfaces in $\mathbb{R}^3$ under the Euclidean group of proper motions.

This example is presented in Cartan [C, 1937], pp. 218-230, but we include it here because its great familiarity to most readers makes it a valuable prototype for understanding the general theory of Chapter I.

The following classification scheme is used for all five examples. Aclass of surfaces will be denoted by a symbol such as I c II a iii , where the upper case Roman numerals refer to the order, the lower case Roman letters itemize the orbit types at that order (beginning at the most singular orbits, which are generally the most interesting geometrically, and ending with the principal orbits), and the lower case Roman numerals list the possible cases of how many and which invariants are independent, (what we shall refer to as the fine structure of that type).

In this example there are four types at the second order, namely, Types: IIa (planes), IIb (spheres), IIc (minimal surfaces), and IId (generic case). Further analysis subdivides the last type into IIdIIIa (right circular cylinders), and IIdIIIb.

Now $M = \mathbb{R}^3$, S is a connected 2-manifold, $G = \mathbb{E}(3)$, o is the origin in $\mathbb{R}^3$, and $f: S \to M$ is a smooth imbedding.

We represent $\mathbb{E}(3)$ by $\{(A, x): A \in SO(3), x \in \mathbb{R}^3\}$, and its action on $\mathbb{R}^3$ is given by $(A, x)y = Ay + x$, for $(A, x) \in \mathbb{E}(3)$, $y \in \mathbb{R}^3$. Then $\mathfrak{g} = \{(X, x): X \in \mathfrak{o}(3), x \in \mathbb{R}^3\}$, $G_0 = \{(A, o): A \in SO(3)\}$ is the isotropy subgroup at o , and its Lie algebra is $\mathfrak{g}_0 = \{(X, o): X \in \mathfrak{o}(3)\}$.

For future reference we note that the composition rule in $\mathbb{E}(3)$ is $(A, x)\cdot(B, y) = (AB, Ay + x)$, the inverse of (A, x) is $(A, x)^{-1} = (A^{-1}, -A^{-1}x)$, and the bracket operation in $\mathfrak{g}$ is $[(X, x), (Y, y)] = ([X, Y], Xy - Yx)$. The adjoint action of G_0 on $\mathfrak{g}$ is $Ad(A, o)(X, x) = (AXA^{-1}, Ax)$. The exponential map $\exp: \mathfrak{g} \to G$ is given by $\exp(X, x) = (e^X, \frac{e^X-1}{X}\cdot x)$, where $e^X = \sum_0^\infty \frac{X^j}{j!}$ and $\frac{e^X-1}{X} = \sum_1^\infty \frac{X^{j-1}}{j!}$.

The Maurer-Cartan form of G is $\Omega_{(A,x)} = (A^{-1}dA, A^{-1}dx) = ((\omega^i_j), (\theta^i))$, where $1 \leq i$, $j \leq 3$, $\omega^i_j + \omega^j_i = 0$, $\omega^i_j = \sum_1^3 A^k_i dA^k_j$, and $\theta^i = \sum_1^3 A^k_i dx^k$, and A^i_j is the ij^{th} entry of $A \in SO(3)$, x^i is the ith row entry in the 3×1 column vector $x \in \mathbb{R}^3$. The Maurer-Cartan equations $d\Omega = -\frac{1}{2}[\Omega, \Omega]$ are $d\omega^i_j = -\sum_1^3 \omega^i_k \wedge \omega^k_j$ and $d\theta^i = -\sum_1^3 \omega^i_j \wedge \theta^j$.

As a vector subspace of $\mathfrak{g}$ complementary to $\mathfrak{g}_0$ we take the $Ad(G_0)$-invariant subspace $\mathfrak{m}_0 = \{(0, x): x \in \mathbb{R}^3\}$, and for a basis of $\mathfrak{m}_0$ we choose $E_i = (0, \varepsilon_i)$, where ε_i is the ith standard basis vector of $\mathbb{R}^3$, $i = 1, 2, 3$. Then G is imbedded in the bundle of frames $L(M)$ of M by the map $h: G \to L(M)$ given by $h(A, x) = (A, x)_*\pi_*(E_1, \ldots, E_3)$, the orthonormal frame at $x \in M$ given by the three columns of $A \in SO(3)$.

The Grassmann bundle of 2-planes on M can be represented as $G_{3,2}(M) = G\times_{G_0} G_{3,2}$, where G_0 acts on G by right multiplication and on $G_{3,2}$ via the representation $\rho_0: G_0 \to GL(3; \mathbb{R})$ given by the adjoint representation of G_0 on $\mathfrak{m}_0$ with respect to the basis E_1, E_2, E_3. In this case $\rho_0(A, o) = A \in SO(3) \subseteq GL(3; \mathbb{R})$, for any $(A, o) \in G_0$.

The zeroth order frames on f are $L_0 = \{(A, x): x = f(s), \text{ some } s \in S\} = \{(A, f(s)): s \in S, A \in SO(3)\}$. $L_0 \to S$ is a principal G_0-bundle over S. The smooth map $\lambda_0: L_0 \to G_{3,2}$ is given by $\lambda_0(A, f(s)) = [h(A, f(s))]^{-1} f_* S_s$, which is a 2-plane in $\mathbb{R}^3$.

The action of G_0 on $G_{3,2}$ is transitive, so for a smooth local cross section of this action we take $W_1 = \left\{ x_0 \equiv \begin{bmatrix} 1 & 0 \\ 0 & 1 \\ 0 & 0 \end{bmatrix} \right\}$, i.e., a single point in $G_{3,2}$, namely the plane in $\mathbb{R}^3$ spanned by ε_1 and ε_2. The first order frames on f with respect to W_1 are the elements of $L_1 = \lambda_0^{-1} W_1$. Any smoothly imbedded surface in M is (locally) the type of W_1. Since $\dim W_1 = 0$, there are no first order invariants. Any two imbedded surfaces in $\mathbb{R}^3$ have at least first order $\mathbb{E}(3)$-contact at any two points.

Geometrically, a first order frame on f at $s \in S$ is a zeroth order frame $(A, f(s))$ such that $f_* S_s$ is the plane in $M_{f(0)}$ spanned by the first two vectors of the frame $h(A, f(s))$.

With respect to the decomposition $\mathfrak{g} = \mathfrak{g}_0 + \mathfrak{m}_0$, we have $\Omega = \omega_0 + \Theta_0$ where $\omega_0 = ((\omega^i_j), 0)$ and $\Theta_0 = (0, (\theta^i))$. If u is a zeroth order frame field along f, and if $hu = (X_1, X_2, X_3; f)$, then $df = \sum_1^3 (u^* \theta^i) \otimes X_i$. Let ϕ^1, ϕ^2 be a coframe field on S, and set $u^* \theta^i = \sum_1^2 x^i_\alpha \phi^\alpha$, for some smooth functions x^i_α on S. Then

$$\lambda_0 \circ u = \begin{bmatrix} x_1^1 & x_2^1 \\ x_1^2 & x_2^2 \\ x_1^3 & x_2^3 \end{bmatrix}.$$ Hence u is a first order frame field along f iff $u^*\theta^3 = 0$.

The isotropy subgroup of G_0 at any point of W_1 is G_1

$$= \{(A, o) \in G_0 : \rho_0(A, o)x_0 = x_0\} = \left\{ \left(\begin{pmatrix} a & \begin{matrix} 0 \\ 0 \end{matrix} \\ \begin{matrix} 0 & 0 \end{matrix} & \varepsilon \end{pmatrix}, o \right) : a \in O(2) \ , \ \varepsilon = \det a \right\}$$

Then $L_1 \to S$ is a principal G_1-bundle, and there are two orientation classes of first order frames, corresponding to the two connected components of G_1.

The $Ad(G_1)$-invariant subspace of $\mathcal{G}_0$ complementary to $\mathcal{G}_1$

$$= \left\{ \left(\begin{pmatrix} 0 & -t & 0 \\ t & 0 & 0 \\ 0 & 0 & 0 \end{pmatrix}, o \right) : t \in \mathbb{R} \right\}$$ is $$\mathfrak{m}_1 = \left\{ \left(\begin{pmatrix} 0 & 0 & -r \\ 0 & 0 & -t \\ r & t & 0 \end{pmatrix}, o \right) : r,t \in \mathbb{R} \right\},$$ and

for a basis of it we take $$E_4 = \left(\begin{pmatrix} 0 & 0 & -1 \\ 0 & 0 & 0 \\ 1 & 0 & 0 \end{pmatrix}, o \right), \quad E_5 = \left(\begin{pmatrix} 0 & 0 & 0 \\ 0 & 0 & -1 \\ 0 & 1 & 0 \end{pmatrix}, o \right).$$

The representation $\rho_1 : G_1 \to GL(m_1; \mathbb{R})$, where $m_1 = 5$, given by the adjoint action of G_1 on $\mathfrak{m}_0 + \mathfrak{m}_1$ with respect to the basis

$E_1, \ldots, E_5$ is $$\rho_1 \left(\begin{pmatrix} a & \begin{matrix} 0 \\ 0 \end{matrix} \\ \begin{matrix} 0 & 0 \end{matrix} & \varepsilon \end{pmatrix}, o \right) = \begin{pmatrix} a & \begin{matrix} 0 \\ 0 \end{matrix} & 0 \\ \begin{matrix} 0 & 0 \end{matrix} & \varepsilon & \begin{matrix} 0 & 0 \end{matrix} \\ 0 & \begin{matrix} 0 \\ 0 \end{matrix} & \varepsilon a \end{pmatrix}, \quad a \in O(2),$$

$\varepsilon = \det a$. We seek an expression for $\lambda_1 : L_1 \to G_{5,2}$, (cf. section I.7).

With respect to the decomposition $\mathcal{G}_0 = \mathcal{G}_1 + \mathfrak{m}_1$, we have

$\omega_0 = \omega_1 + \Theta_1$, where $\omega_1 = \left(\begin{pmatrix} 0 & \omega_2^1 & 0 \\ \omega_1^2 & 0 & 0 \\ 0 & 0 & 0 \end{pmatrix}, 0\right)$ and

$\Theta_1 = \left(\begin{pmatrix} 0 & 0 & \omega_3^1 \\ 0 & 0 & \omega_3^2 \\ \omega_1^3 & \omega_2^3 & 0 \end{pmatrix}, 0\right)$. If u is a first order frame field along f , then $u^*\theta^3 = 0$, and $\phi^1 = u^*\theta^1$, $\phi^2 = u^*\theta^2$ is a coframe field on S , so $u^*\omega_1^3 = x\phi^1 + y\phi^2$ and $u^*\omega_2^3 = y\phi^1 + z\phi^2$, for some smooth functions $x, y, \tilde{y}, z$ on S . Then $\lambda_1 \circ u = \begin{bmatrix} 1 & 0 \\ 0 & 1 \\ 0 & 0 \\ x & y \\ \tilde{y} & z \end{bmatrix} : S \to G_{5,2}$.

If we take the exterior derivative of both sides of $u^*\theta^3 = 0$, (for a first order frame field u along f) , and apply the Maurer-Cartan equations of G , we find that $y = \tilde{y}$. Hence

$$\lambda_1(L_1) \subseteq \left\{\begin{bmatrix} I \\ 0\ 0 \\ X \end{bmatrix} : X = 2 \times 2 \text{ symmetric matrix}\right\} , \quad I = 2 \times 2 \text{ identity}$$

matrix.

We need to find a cross section of the action (G_1, ρ_1) , which is $\rho_1\left(\left(\begin{pmatrix} a & \begin{matrix} 0 \\ 0 \end{matrix} \\ 00 & \varepsilon \end{pmatrix}, 0\right)\right)\begin{bmatrix} I \\ 0\ 0 \\ X \end{bmatrix} = \begin{bmatrix} a \\ 0\ 0 \\ \varepsilon\ a\ X \end{bmatrix} = \begin{bmatrix} I \\ 0\ 0 \\ \varepsilon\ a\ Xa^{-1} \end{bmatrix}$, i.e. of the action G_1 on the space $\mathcal{S}$ of all 2×2 symmetric matrices given by $X \mapsto \varepsilon a X a^{-1}$, for $a \in O(2)$, $\varepsilon = \det a$.

A cross section of this action is

$$\left\{\begin{bmatrix} r & 0 \\ 0 & t \end{bmatrix} : t \geqslant 0 \ , \ |r| \leqslant t\right\}.$$ There are four orbit types:

Type IIa: those orbits met by $W_{2a} = \left\{\begin{bmatrix} I \\ 0 \end{bmatrix}\right\}$;

Type IIb: those orbits met by $W_{2b} = \left\{\begin{bmatrix} I \\ 0 \ 0 \\ rI \end{bmatrix} : r > 0\right\}$;

Type IIc: those orbits met by $W_{2c} = \left\{\begin{bmatrix} I & \\ 0 & 0 \\ -r & 0 \\ 0 & r \end{bmatrix} : r > 0\right\}$; and

Type IId: those orbits met by $W_{2d} = \left\{\begin{bmatrix} I & \\ 0 & 0 \\ r & 0 \\ 0 & t \end{bmatrix} : |r| < t \ , \ t > 0\right\}$.

Type IIa surfaces. Suppose $f: S \to M$ is Type IIa; i.e., it is the type of W_{2a} (cf. section I.6); i.e., $\lambda_1(L_1) \subseteq \rho_1(G_1) \cdot W_{2a} = W_{2a}$. Thus, for any first order frame field u along f , $u^*\omega_1^3 = 0 = u^*\omega_2^3$. The second order frames on S are the elements of $L_2 = \lambda_1^{-1} W_{2a} = L_1$. Since $\dim W_{2a} = 0$, there are no second order invariants.

The action of (G_1, ρ_1) on W_{2a} is trivial, and there are no non-constant invariants. Consequently all higher order frames are the same as the first order frames, and $f(S)$ is an open submanifold of a homogeneous submanifold, by the Corollary of section I.12.

In fact, let $\mathfrak{h}$ be the involutive, 3-dimensional, left-invariant distribution on G defined by the equations $\{\theta^3 = 0 , \ \omega_1^3 = 0 , \ \omega_2^3 = 0\}$ The analytic subgroup of G with Lie algebra $\mathfrak{h}$ is

$$H = \left\{ \left(\begin{pmatrix} a & & 0 \\ & & 0 \\ 0 & 0 & 1 \end{pmatrix}, \begin{pmatrix} x \\ \\ 0 \end{pmatrix} \right) : a \in SO(2) \ , \ x \in \mathbb{R}^2 \right\} ,$$ which is isomorphic to $\mathbb{E}(2)$. By the Corollary of I.12, $f(S)$ is G-congruent to an open submanifold of $\pi H = \mathbb{R}^2 \subseteq \mathbb{R}^3$, the $\varepsilon_1\varepsilon_2$-plane in $\mathbb{R}^3$.

Type IIb surfaces.

Suppose $f: S \to M$ is Type IIb; i.e., it is the type of W_{2b}.

Then $\lambda_1(L_1) \subseteq \rho_1(G_1)W_{2b} = \left\{ \begin{bmatrix} I \\ 0 \ 0 \\ r \ I \end{bmatrix} : r \neq 0 \right\}$. Thus, for any first order frame field u along f, $u^*\omega_1^3 = k\phi^1$, $u^*\omega_2^3 = k\phi^2$ for some non-vanishing smooth function k on S.

Lemma. k is constant.

Proof: Take the exterior derivative of the above two equations, and apply the Maurer-Cartan equations of G. In fact, $dk = k_{;1}\phi^1 + k_{;2}\phi^2$ and $u^*\omega_1^2 = k^3\phi^1 + k^4\phi^2$ for some smooth functions $k_{;1}$, $k_{;2}$, k^3, and k^4 on S. Now $du^*\omega_1^3 = -u^* \sum_j \omega_j^3 \wedge \omega_1^j = kk^3\phi^1 \wedge \phi^2$, and $d(k\phi^1)$ $= -k_{;2}\phi^1 \wedge \phi^2 - ku^* \sum_j \omega_j^1 \wedge \phi^j = (kk^3 - k_{;2})\phi^1 \wedge \phi^2$. Thus $kk^3 = kk^3 - k_{;2}$ and $k_{;2} = 0$. Similarly, from the equation $u^*\omega_2^3 = k\phi^2$ it follows that $k_{;1} = 0$. Hence $dk = 0$, and k is constant on S.

The second order frames on f are the elements of $L_2 = \lambda_1^{-1}W_{2a}$, which is just one of the two orientation classes of first order frames on f. $L_2 \to S$ is a principal G_2-bundle over S, where G_2 is the connected component of the identity of G_1. Its Lie algebra, $\mathfrak{g}_2$, equals $\mathfrak{g}_1$, and since there are no non-constant invariants, it follows that all higher order frames are the same; i.e., $L = L_2$ for $j \geq 2$. A second order frame field along f is a first order frame field u for which $k > 0$.

Let $\mathfrak{h}$ be the involutive, 3-dimensional, left-invariant distribution on G defined by the equations $\{\theta^3 = 0, \ \omega_1^3 = k\theta^1, \ \omega_2^3 = k\theta^2\}$, where k is the positive constant second order invariant.

If H denotes the analytic subgroup of G whose Lie algebra is $\mathfrak{h}$, then it requires an elementary computation to show that $\pi H = \{(x, y, z) \in \mathbb{R}^3 : x^2 + y^2 + (z + r)^2 = r^2\}$, where $r = k^{-1}$; i.e.- πH is the sphere in $\mathbb{R}^3$ of radius r centered at $(0, 0, -r)$. By the Corollary in Section I.12, $f(S)$ is G-congruent to an open submanifold of this sphere.

Types IIc and IId.

Suppose $f: S \to M$ is Type IIc; i.e., it is the type of W_{2c}. Then $\lambda_1(L_1) \subseteq \rho_1(G_1)W_{2c}$, and the second order frames on f are the elements of $L_2 = \lambda_1^{-1}W_{2c}$. The isotropy subgroup of G_1 at any point of W_{2c} is $G_2 = \left\{ \begin{pmatrix} \pm I & 0 \\ 0 & 1 \end{pmatrix}, \begin{pmatrix} \pm K & 0 \\ 0 & -1 \end{pmatrix} : K = \begin{pmatrix} 0 & 1 \\ 1 & 0 \end{pmatrix} \right\}$, and $L_2 \to S$ is a principal G_2-bundle.

A second order frame field along f is any first order frame field u for which $u^*\omega_1^3 = k^1\phi^1$, $u^*\omega_2^3 = k^2\phi^2$, where k^1 and k^2 are smooth functions on S and $-k^1 = k^2 > 0$, (and $\phi^i = u^*\theta^i$, $i = 1, 2$) . The function k^2 is the second order invariant of f .

Suppose $f\colon S \to M$ is Type IId; i.e., it is the type of W_{2d} . Then $\lambda_1(L_1) \subseteq \rho_1(G_1)W_{2d}$, and the second order frames on f are the elements of $L_2 = \lambda_1^{-1}W_{2d}$. The isotropy subgroup of G_1 at any point of W_{2d} is $G_2 = \left\{ \begin{pmatrix} \pm I & 0 \\ 0 & 1 \end{pmatrix} \right\}$, and $L_2 \to S$ is a principal G_2-bundle.

A second order frame field along f is any first order frame field u for which $u^*\omega_1^3 = k^1\phi^1$, $u^*\omega_2^3 = k^2\phi^2$, where k^1 and k^2 are smooth functions on S and $|k^1| < k^2$, $k^2 > 0$. They are the second order invariants of f .

From this point on we discuss Types IIc and IId together. For either case the Lie algebra of G_2 is $\mathfrak{g}_2 = (0)$. The $\mathrm{Ad}(G_2)$-invariant subspace of $\mathfrak{g}_1$ complementary to $\mathfrak{g}_2$ is, of course $\mathfrak{m}_2 = \mathfrak{g}_1$ $= \left\{ \left(\begin{pmatrix} 0 & -t & 0 \\ t & 0 & 0 \\ 0 & 0 & 0 \end{pmatrix}, 0 \right) : t \in \mathbb{R} \right\}$, and for a basis of $\mathfrak{m}_2$ we take

$E_6 = \left\{\begin{pmatrix} 0 & -1 & 0 \\ 1 & 0 & 0 \\ 0 & 0 & 0 \end{pmatrix}, 0\right\}$. Then for $\begin{pmatrix} a & & 0 \\ & & 0 \\ 0 & 0 & \varepsilon \end{pmatrix} \in G_2$, $\rho_2\begin{pmatrix} a & \\ & \varepsilon \end{pmatrix}$

$$= \begin{pmatrix} a & & & & & \\ & \varepsilon & & & & \\ & & \varepsilon a & & & \\ & & & \varepsilon & & \\ & & & & 1 & \\ & & & & & 1 \end{pmatrix} \in GL(8;\mathbb{R})$$

, and its action on $\psi_2^{-1}W_{2c}$ (or $\psi_2^{-1}W_{2d}$), which is

$$\left\{\begin{bmatrix} I \\ \begin{matrix} 0 & 0 \\ x^1 & 0 \\ 0 & x^2 \\ x^3 & x^4 \end{matrix} \\ X \end{bmatrix} : X \in \mathbb{R}^{4\times 4},\ x^1, \ldots, x^4 \in \mathbb{R}\right\},$$

is given by

$$\rho_2\begin{pmatrix} a & \\ & \varepsilon \end{pmatrix}\begin{bmatrix} I \\ \begin{matrix} 0 & 0 \\ x^1 & 0 \\ 0 & x^2 \\ x^3 & x^4 \end{matrix} \\ X \end{bmatrix} = \begin{bmatrix} I \\ \begin{matrix} 0 & 0 \\ x^1 & 0 \\ 0 & x^2 \end{matrix} \\ \varepsilon(x^3, x^4)a^{-1} \\ Xa^{-1} \end{bmatrix}.$$

There are two orbit types, namely the set where $x^3 = x^4 = 0$ and $X = 0$, and the complement of this set. A local cross section for the latter is given by an open half space or quarter space of $\mathbb{R}^6$, as the case may be.

With respect to the decomposition $\mathfrak{g}_1 = \mathfrak{g}_2 + \hat{\mathfrak{m}}_2$, we have $\omega_1 = \omega_2 + \Theta_2$, where $\omega_2 = 0$ and $\Theta_2 = \omega_1 = \left\{\begin{pmatrix} 0 & \omega_2^1 & 0 \\ \omega_1^2 & 0 & 0 \\ 0 & 0 & 0 \end{pmatrix}, 0\right\}$. If u is a second order frame field along f , with $\phi^i = u^*\theta^i$, $i = 1, 2$, then $u^*\omega_1^3 = k^1\phi^1$, $u^*\omega_2^3 = k^2\phi^2$, where k^1 and k^2 are the second order invariants of f , $k^2 \geqslant 0$, and $k^1 = -k^2$ in Type IIc ,

$|k^1| < k^2$ in Type IId. We set $u^*\omega_1^2 = k^3\phi^1 + k^4\phi^2$, $dk^i = k^i_{;1}\phi^1 + k^i_{;2}\phi^2$, $i = 1, 2$. Then

$$\lambda_2 \circ u = \begin{bmatrix} 1 & 0 \\ 0 & 1 \\ 0 & 0 \\ k^1 & 0 \\ 0 & k^2 \\ k^3 & k^4 \\ k^1_{;1} & k^1_{;2} \\ k^2_{;1} & k^2_{;2} \end{bmatrix} : S \to G_{8,2} .$$

Taking the exterior derivative of these equations, and applying the Maurer-Cartan equations of G, we get the structure equations of f:

$$\begin{aligned} d\phi^1 &= k^3\phi^1 \wedge \phi^2 , \qquad d\phi^2 = k^4\phi^1 \wedge \phi^2 \\ k^3 &= k^1_{;2}(k^1 - k^2)^{-1} , \qquad k^4 = k^2_{;1}(k^1 - k^2)^{-1} \\ k^1k^2 &= (k^3)^2 + (k^4)^2 + k^3_{;2} - k^4_{;1} . \end{aligned} \tag{1}$$

Corresponding to the two orbit types of (G_2, ρ_2) we define surfaces of Types IIcIII, IIdIIIa, and IIdIIIb. (By equations (1), $\lambda_2 \circ u$ cannot take all values in the singular orbit of ρ_2 when f is of Type IIc.)

Type IIdIIIa.

We say $f: S \to M$ is of this type if it is Type IId and k^3, k^4, dk^1, dk^2 are all identically zero (with respect to some, hence any, second order frame field u along f). By (1) this is equivalent to k^1 and k^2 being constant, and then by the last equation in (1), $k^1 = 0 < k^2$. f has no non-constant invariants.

Let $\mathcal{J}_k$ be the involutive, 2-dimensional, left-invariant distribution defined on G by the equations $\{\theta^3 = 0, \omega_1^3 = 0, \omega_2^3 = k\theta^2$ $\omega_1^2 = 0\}$. Thus $\mathcal{J}_k = \left\{ \left(\begin{pmatrix} 0 & 0 & 0 \\ 0 & 0 & kr \\ 0 & kr & 0 \end{pmatrix}, \begin{pmatrix} t \\ r \\ 0 \end{pmatrix} \right) : r,t \in \mathbb{R} \right\}$, where k is a positive constant. Its analytic subgroup in G is

$$H_k = \left\{ \left(\begin{pmatrix} 1 & 0 & 0 \\ 0 & \cos kr & -\sin kr \\ 0 & \sin kr & \cos kr \end{pmatrix}, \begin{pmatrix} t \\ k^{-1} \sin kr \\ k^{-1}(1 - \cos kr) \end{pmatrix} \right) : t, r \in \mathbb{R} \right\}.$$

Hence, by the Corollary of section I.12, $f(S)$ is G-congruent to an open submanifold of $\pi H_{k^2} = \{(x, y,z) \in \mathbb{R}: y^2 + (z - R)^2 = R^2\}$, where $R = (k^2)^{-1}$; i.e., a right circular cylinder of radius $(k^2)^{-1}$.

Type IIdIIIb.

We say $f: S \to M$ is of this type if it is of Type IId and $\lambda_2 \circ u(S)$ is contained in the union of the principal orbits of (G_2, ρ_2) for some, hence any, second order frame field u along f. In other words, at least one of k^3, k^4, dk^1, and dk^2 is nonzero at each point of S. By (1), this is equivalent to one of the functions k^1 , k^2 being non-constant. We shall discuss the case when k^1 is non-constant, say $k^1_{;1} \neq 0$. Then we define the third order frame field along f to be the second order frame field with respect to which $k^1_{;1} > 0$. Thus L_3 is just one of the two connected components of L_2 , and the third order frame at each point is uniquely determined. Clearly $L_j = L_3$ for all $j \geqslant 3$.

The six functions k^3 , k^4 , $k^1_{;1}$, $k^1_{;2}$, $k^2_{;1}$, $k^2_{;2}$ are the third order invariants of f . Equations (1) express k^3 and k^4 in terms of $k^1_{;2}$, $k^2_{;1}$, k^1 and k^2 . All higher order invariants of f are just derivatives of these four functions. For example, the fourth order invariants are $k^i_{;jk}$, where $1 \leqslant i, j , k \leqslant 2$, defined by $dk^i_{;j} = k^i_{;j1}\phi^1 + k^i_{;j2}\phi^2$.

Finally, the determination of the Frenet frame of f , and the completion of the classification, depends on how many, and which, of the invariants are independent. There are three cases:

(i) k^1 and k^2 are independent, i.e. $dk^1 \wedge dk^2 \neq 0$ on S ;

(ii) k^1 and k^2 are functionally dependent, but one of them and one of $\{k^i_{;j} : 1 \leqslant i , j \leqslant 2\}$ are independent.

(iii) k^1 , $k^2_{;1}$, $k^2_{;2}$, (and hence all other invariants) are functions of k^2 . (Or k^2 is constant and all invariants are functions of k^1 .)

Type IIdIIIbi.

We say f is of this type if it is of Type IIdIIIb and case (i) above holds. Taking k^1, k^2 as local coordinates on S , it follows that all higher order invariants are functions of k^1, k^2 . Hence the Frenet frames of f are of order $q = 2$, which means at least third order contact is required for congruence.

Let $k^i_{;j} = h_{ij}(k^1, k^2)$, $1 \leq i$, $j \leq 2$, where the h_{ij} are smooth functions of two variables. Then, by Theorem 3 of section I.11, f is G-congruent to $\tilde{f}: \tilde{S} \to M$ iff $\tilde{f}$ is also of Type IIdIIIb; and $\tilde{h}_{ij} = h_{ij}$, where we denote all quantities for $\tilde{f}$ by a tilde over the same letter as for f .

Note there is no Type IIcIII surface for which case (i) holds.

Type IIdIIIbii and Type IIcIIIii.

We say f is of one of these types if it is of Type IIcIII or Type IIdIIIb and case (ii) above holds. The Frenet frames have order $q = 3$.

Type IIdIIIbiii and Type IIcIIIiii.

We say f is of one of these types if it is of Type IIcIII or Type IIdIIIb and case (iii) above holds. The Frenet frames have order $q = 2$. For such surfaces which are real analytic, there is a 1-parameter subgroup H of G which sends $f(S)$ into itself. $f(S)$ consists of the H-orbits of the points on a profile curve, and this is either a cylinder (orbits are lines), or a surface of revolution (orbits are circles), or a generalized helicoid (orbits are helices).

III. Curves in real Grassmannians.

To simplify the exposition we consider the case of real curves in $\tilde{G}_{4,2}$, the real Grassmann manifold of oriented 2-planes in $\mathbb{R}^4$, under the action of $G = SO(4)$. For curves in $\tilde{G}_{n,q}$, for $1 \leqslant q \leqslant [\frac{n}{2}]$, the analysis is very similar, but the linear algebra becomes more cumbersome. This example is considered in Griffiths [G, 1974].

In this example there are three types at the first order, called Types Ia, Ib and Ic. The second type is discussed no further, and further analysis of the other two leads to the Types: IaIIa, IaIIb, IcIIa and IcIIb.

Now $M = \tilde{G}_{4,2}$, which we realize as $\mathbb{R}^{4\times 2*}/GL^+(2,\mathbb{R})$, where $\mathbb{R}^{4\times 2}$ denotes the space of 4×2 real matrices, $\mathbb{R}^{4\times 2*}$ is the open subset of 4×2 real matrices of rank 2, and $GL^+(2,\mathbb{R})$ denotes the 2×2 non-singular matrices with positive determinant. The latter acts on $\mathbb{R}^{4\times 2*}$ by ordinary matrix multiplication from the right. If $X \in \mathbb{R}^{4\times 2*}$, we let $[X]$ denote its equivalence class in $\tilde{G}_{4,2}$.

The group $G = SO(4)$ acts on M by ordinary matrix multiplication from the left. For the origin of M we take $o = \begin{bmatrix} 1 & 0 \\ 0 & 1 \\ 0 & 0 \\ 0 & 0 \end{bmatrix}$, and then the isotropy subgroup of G at o is $G_0 = SO(2) \times SO(2)$

$$= \left\{ \begin{pmatrix} a & 0 \\ 0 & b \end{pmatrix} \in G\colon\ a,\ b \in SO(2) \right\}.$$

The Lie algebra of G is $\mathfrak{g} = \mathcal{O}(4)$, and of G_o is $\mathfrak{g}_o$ $= \left\{ \begin{pmatrix} X & o \\ o & Y \end{pmatrix} \in \mathfrak{g} : X, Y \in \mathcal{O}(2) \right\}$. The Maurer-Cartan form of G is Ω_A $= A^{-1}dA = (\omega^i_j)$, where $1 \leqslant i$, $j \leqslant 4$, $\omega^i_j + \omega^j_i = 0$, ω^i_j $= \sum_1^4 A^k_i dA^k_j$, and A^i_j is the ij^{th} entry of $A \in SO(4)$. The Maurer-Cartan equations $d\Omega = -\frac{1}{2}[\Omega, \Omega]$ are $d\omega^i_j = -\sum_1^4 \omega^i_k \wedge \omega^k_j$.

As a vector subspace of $\mathfrak{g}$ complementary to $\mathfrak{g}_o$ we take the $Ad(G_o)$-invariant subspace $\mathfrak{m}_o = \left\{ \begin{pmatrix} o & -X \\ X & o \end{pmatrix} \in \mathfrak{g} : X \in \mathbb{R}^{2\times 2} \right\}$, and for a basis of $\mathfrak{m}_o$ we choose $E_1 = \begin{pmatrix} 0 & \begin{matrix} -1 & o \\ o & o \end{matrix} \\ \begin{matrix} 1 & o \\ o & o \end{matrix} & 0 \end{pmatrix}$, ..., $E_4 = \begin{pmatrix} 0 & \begin{matrix} o & o \\ o & -1 \end{matrix} \\ \begin{matrix} o & o \\ o & 1 \end{matrix} & \end{pmatrix}$. Then G is imbedded in the bundle of frames $L(M)$ of M by the map $h: G \to L(M)$ given by $hA = A_* \pi_*(E_1, \ldots, E_4)$, where $\pi: G \to G/G_o = M$ is the natural projection.

We consider a smooth imbedding $f: S \to M$, where S is a connected 1-manifold. The zeroth order frames on f are elements of L_0 $= \{A \in G: \pi(A) \in f(S)\}$, a G_0-bundle over S . The smooth map $\lambda_0: L_0 \to G_{4,1} = \mathbb{RP}^3$ is given by $\lambda_0(A) = [hA]^{-1} f_* S_s$, which is a 1-plane in $\mathbb{R}^4$, where $f(s) = \pi(A)$.

The action of G_0 on $G_{4,1}$ is given by the representation $\rho_0: G_0 \to GL(4; \mathbb{R})$ given by the adjoint representation of G_0 on $\mathfrak{m}_0$ with respect to the basis $E_1, \ldots, E_4$. If we denote $\mathbb{R}^4$ by $\mathbb{R}^{2\times 2}$, then for $(a, b) \in G_0 = SO(2) \times SO(2)$ and $X \in \mathbb{R}^{2\times 2}$, we have

$\rho_0(a, b)X = bXa^{-1}$. If we denote the elements of $G_{4,1} = \mathbb{R}^{2\times 2*}/\mathbb{R}^*$ by $\langle X\rangle$, then a cross section of this action on $G_{4,1}$ is given by the set $\left\{\begin{pmatrix}1 & 0\\ 0 & r\end{pmatrix}: |r| \leq 1\right\}$.

There are three orbit types:

Type Ia – those orbits meeting $\left\langle\begin{matrix}1 & 0\\ 0 & 1\end{matrix}\right\rangle$;

Type Ib – those orbits meeting $\left\langle\begin{matrix}1 & 0\\ 0 & -1\end{matrix}\right\rangle$; and

Type Ic – those orbits meeting $\left\{\left\langle\begin{matrix}1 & 0\\ 0 & r\end{matrix}\right\rangle: |r| < 1\right\}$.

With respect to the decomposition $\mathfrak{g} = \mathfrak{g}_0 + \mathfrak{m}_0$, we have $\Omega = \omega_0 + \Theta_0$, where ω_0 is the $\mathfrak{g}_0$-component and $\Theta_0 = \omega_1^3 \otimes E_1 + \omega_2^3 \otimes E_2 + \omega_1^4 \otimes E_3 + \omega_2^4 \equiv E_4 \otimes \sum_1^4 \theta^i \otimes E_i$. Let ϕ be a coframe field on S . For a zeroth order frame field u along f , set $u^*\theta^i = x^i\phi$, where x^i is a smooth function on S , $1 \leq i \leq 4$. Then $\lambda_0 \circ u = \left\langle\begin{matrix}x^1 & x^2\\ x^3 & x^4\end{matrix}\right\rangle: S \to G_{4,1}$.

<u>Type Ia curves.</u>

We say that $f: S \to M$ is of Type Ia if it is the type of the smooth local cross section $W_1 = \left\{\left\langle\begin{matrix}1 & 0\\ 0 & 1\end{matrix}\right\rangle\right\}$ of ρ_0 ; i.e., there exists a zeroth order frame field u along f such that $u^*\omega_2^3 = 0 = u^*\omega_1^4$ and $u^*\omega_2^4 = u^*\omega_1^3$. There are no first order invariants.

The isotropy subgroup of G_0 at $\left\langle\begin{matrix}1 & 0\\ 0 & 1\end{matrix}\right\rangle$ is $G_1 = \left\{\begin{bmatrix}a & 0\\ 0 & \pm a\end{bmatrix} \in G_0: a \in SO(2)\right\}$, which has two connected components, corresponding to two orientation classes of first order frames on f .

The $\mathrm{Ad}(G_1)$-invariant subspace of $\mathfrak{g}_0$ complementary to $\mathfrak{g}_1$ $= \left\{ \begin{pmatrix} X & 0 \\ 0 & X \end{pmatrix} : X \in \mathfrak{o}(2) \right\}$ is $\mathfrak{m}_1 = \left\{ \begin{pmatrix} X & 0 \\ 0 & -X \end{pmatrix} : X \in \mathfrak{o}(2) \right\}$, and for a basis of it we take $E_5 = \begin{pmatrix} J & 0 \\ 0 & -J \end{pmatrix}$, where $J = \begin{pmatrix} 0 & -1 \\ 1 & 0 \end{pmatrix}$ The representation $\rho_1 : G_1 \to GL(m_1; \mathbb{R})$, $m_1 = 5$, given by the adjoint action of G_1 on $\mathfrak{m}_0 + \mathfrak{m}_1$ with respect to the basis $E_1, \ldots, E_5$ is $\rho_1 \begin{pmatrix} a & 0 \\ 0 & \pm a \end{pmatrix}$ $= \begin{pmatrix} \pm I_4 & 0 \\ 0 & 1 \end{pmatrix}$, where I_n is the $n \times n$ identity matrix.

With respect to $\mathfrak{g}_0 = \mathfrak{g}_1 + \mathfrak{m}_1$, we have $\omega_0 = \omega_1 + \Theta_1$, where ω_1 is the $\mathfrak{g}_1$-component and $\Theta_1 = \frac{1}{2}(\omega_1^2 - \omega_3^4) \otimes E_5$. If u is a first order frame field along f, then $u^*\omega_1^3 = u^*\omega_2^4 = dt$, for some coordinate function t on S and we set $u^*\frac{1}{2}(\omega_1^2 - \omega_3^4) = k^1 dt$, where k^1 is a smooth function on S. Then the map $\lambda_1 : L_1 \to G_{5,1}$ is given by $\lambda_1 \circ u = {}^t[1, 0, 0, 1, k^1]$. The action of G_1, ρ_1 on $\lambda_1(L_1)$ has two orbit types: namely $k^1 = 0$ and $k^1 \neq 0$.

We say f is of <u>Type</u> IaIIa if $k^1 = 0$ with respect to some, hence any, first order frame field along f. The isotropy subgroup of G_1 at ${}^t[1, 0, 0, 1, 0]$ is $G_2 = G_1$, so $L_j = L_1$ for all $j \geq 2$. There are no invariants and the Frenet frames are the first order frames $L_1 \to S$, a G_1-bundle. By the Corollary in section I.12, $f(S)$ is G-congruent to an open submanifold of the homogeneous submanifold $\pi(H)$, where H is the analytic subgroup of G whose Lie algebra is $\mathfrak{h} = \{\omega_2^3 = 0\ ,\ \omega_1^4 = 0\ ,\ \omega_2^4 = \omega_1^3\ ,\ \omega_1^2 = \omega_3^4\}$.

We say f is of Type IaIIb if $k^1 \neq 0$, with respect to some, hence any, first order frame field along f . Then $k^1 > 0$ with respect to some first order frame field, and this condition characterizes second order frames. The isotropy subgroup of G_1 at ${}^t[1, 0, 0, 1, k^1]$, for $k^1 > 0$, is G_2 = the connected component of the identity of G_1 , so L_2 is just one of the two orientation classes of L_1 . Since ρ_1 is trivial on G_2 , and $\mathfrak{g}_2 = \mathfrak{g}_1$ so that $\mathfrak{m}_2 = 0$, it follows that $L_j = L_2$ for all $j \geq 2$. The Frenet frames are first order if k^1 is constant, in which case we call f Type IaIIbi; and they are second order if the second order invariant k^1 is non-constant, in which case we call f Type IaIIbii.

A Type IaIIbi curve $f: S \to M$ has no non-constant invariants, so by the Corollary of section I.12, if $k = k^1$ is the positive constant second order invariant of f , then f(S) is G-congruent to an open submanifold of $\pi(H_k)$, where H_k is the analytic subgroup of G whose Lie algebra is $\mathfrak{h}_k = \{\omega_2^3 = 0\ ,\ \omega_1^4 = 0\ ,\ \omega_2^4 = \omega_1^3\ ,\ \omega_1^2 - \omega_3^4 = 2k\omega_1^3\}$

If k^1 is any positive, smooth function of t , then there exists a smooth curve $f: S \to M$ of Type IaIIbii whose second order invariant is k^1 and whose first order invariant form $u^*\omega_1^3$ is dt . There are no structure equations because dim S = 1 .

Type Ib curves.

We say that $f: S \to M$ is of Type Ib if it is the type of the smooth local cross section $W_1 = \left\{\begin{pmatrix}1 & 0\\ 0 & -1\end{pmatrix}\right\}$ of ρ_0 ; i.e., there exists a zeroth order frame field u along f such that $u^*\omega_2^3 = 0 = u^*\omega_1^4$ and $y^*\omega_2^4 = -u^*\omega_1^3$. There are no first order invariants. Due to the similarity with the Type Ia case, we will not discuss the Type Ib curves any further.

Type Ic curves.

We say that $f: S \to M$ is of Type Ic if it is the type of the smooth local cross section $W_1 = \left\{\begin{pmatrix}1 & 0\\ 0 & r\end{pmatrix}: |r| < 1\right\}$ of ρ_0 ; i.e., there exists a zeroth order frame field u along f such that $u^*\omega_2^3 = 0 = u^*\omega_1^4$ and $u^*\omega_2^4 = k^1u^*\omega_1^4$, where k^1 is a smooth function on S satisfying $|k^1| < 1$. It is the first order invariant of f .

The isotropy subgroup of (G_0, ρ_0) at any point of W_1 is $G_1 = \left\{\begin{pmatrix}a & 0\\ 0 & \pm a\end{pmatrix} \in G_0 : a = \begin{pmatrix}\varepsilon & 0\\ 0 & \delta\end{pmatrix}, \varepsilon, \delta = \pm 1\right\}$, an eight-element group. The first order frames are uniquely determined up to eight orientation classes.

Since $\mathfrak{g}_1 = (0)$, the $Ad(G_1)$-invariant subspace of $\mathfrak{g}_0$ complementary to $\mathfrak{g}_1$ is $\mathfrak{m}_1 = \mathfrak{g}_0$, and for a basis of it we take $E_5 = \begin{pmatrix}J & 0\\ 0 & 0\end{pmatrix}$, $E^6 = \begin{pmatrix}0 & 0\\ 0 & J\end{pmatrix}$, where $J = \begin{pmatrix}0 & -1\\ 1 & 0\end{pmatrix}$. The representation $\rho_1: G_1 \to GL(m_1 + \mu_1, \mathbb{R})$ $m_1 = 6$, $\mu_1 = 1$, is given by $\rho_1\begin{pmatrix}a & 0\\ 0 & \pm a\end{pmatrix} = \begin{pmatrix}\pm I_4 & 0\\ 0 & I_3\end{pmatrix}$.

Since $\mathcal{J}_1 = (0)$, we have $\Theta_1 = \omega_0 = \omega_1^2 \otimes E_5 + \omega_3^4 \otimes E_6$. If u is a first order frame field along f , then $u^*\omega_1^3 = dt$, for some coordinate function t on S , $u^*\omega_2^4 = k^1 dt$, and we set $dk^1 = k^1_{;1}dt$, $u^*\omega_1^2 = k^2 dt$, $u^*\omega_3^4 = k^3\,dt$, where $k^2, k^3, k^1_{;1}$ are smooth functions on S . The map $\lambda_1 : L_1 \to G_{6,1}$ is given by $\lambda_1 \circ u = {}^t[1, 0, 0, k^1, k^2, k^3, k^1_{;1}]$. The action of (G_1, ρ_1) on $k^2k^3k^1_{;1}$-space has two orbit types: the origin $k^2 = k^3 = k^1_{;1} = 0$ and its complement. Any open half-space is a smooth local cross section for the latter.

We say that f is of Type IcIIa if $k^2 = k^3 = k^1_{;1} = 0$ with respect to some, hence any, first order frame field along f . The isotropy subgroup of G_1 at ${}^t[1, 0, 0, k^1, 0, 0, 0]$ is $G_2 = G_1$, so $L_j = L_1$ for all $j \geqslant 2$. Since $k^1_{;1} = 0$, k^1 is constant and there are no non-constant invariants. By the Corollary of I.12, if $k^1 = k$ is the constant first order invariant of f , so $|k| < 1$, then $f(S)$ is G-congruent to an open submanifold of $\pi(H_k)$, where H_k is the analytic subgroup of G whose Lie algebra is $\mathfrak{h}_k = \{\omega_2^3 = 0\ ,\ \omega_1^4 = 0\ ,\ \omega_2^4 = k\omega_1^3\ ,\ \omega_1^2 = 0\ ,\ \omega_3^4 = 0\}$. Notice $\pi(H_k)$ is a geodesic in M , with respect to the canonical (Riemannian) connection on M .

We say that f is of Type IcIIb if at least one of k^2, k^3 , and $k^1_{;1}$ does not vanish. There are three cases:

We say f is of Type IcIIbi if $k^1_{;1} \neq 0$. For a smooth local cross section take $k^1_{;1} > 0$; i.e., this is the condition defining second order frame fields along f . Then G_2

$$= \left\{ \begin{pmatrix} a & 0 \\ 0 & a \end{pmatrix} \in G_1 : a = \begin{pmatrix} \varepsilon & 0 \\ 0 & \delta \end{pmatrix} ,\ \varepsilon, \delta = \pm 1 \right\} ,\ \text{so}\ \mathcal{J}_2 = \mathcal{J}_1 = (0) .$$

Since ρ_j is trivial on $G_j = G_2$ for all $j \geq 2$, it follows that $L_j = L_2$ for all $j \geq 2$. Furthermore, $k^1_{;1} \neq 0$ implies k^1 is non-constant, and hence k^2, k^3 and $k^1_{;1}$ are functions of k^1. The Frenet frames are first order.

We say f is of Type IcIIbii if $k^1_{;1} = 0$ and k^2 and k^3 are constant, not both 0. The Frenet frames are order 1, and there are no non-constant invariants. f(S) is G-congruent to an open submanifold of $\pi(H)$, where H is the analytic subgroup of G whose Lie algebra is

$\mathfrak{h} = \{\omega^3_2 = 0\ ,\ \omega^4_1 = 0\ ,\ \omega^4_2 = k^1\omega^3_1\ ,\ \omega^2_1 = k^2\omega^3_1\ ,\ \omega^4_3 = k^3\omega^3_1\}$. Since one of k^2 and k^3 is non-zero, $\pi(H)$ is not a geodesic in M, but rather it is the analog of a helix.

Finally, we say f is of Type IcIIbiii if $k^1_{;1} = 0$, i.e., k^1 is constant, but at least one of k^2 and k^3 is non-constant. Then the Frenet frames are second order. If, say, k^2 is non-constant, then $u^*\omega^3_1 = m\,dk^2$, $k^1 = \text{constant}$, $k^3 = n(k^2)$ for some functions m and n of a real variable. The functions m and n and the constant k^1 may be arbitrarily specified, since for $\dim S = 1$ there are no structure equations.

IV. Holomorphic curves in complex projective space.

In order to keep the exposition simple and complete, we consider only the case of holomorphic curves in $\mathbb{C}P^2$ under the action of $SU(3)$. The analysis of holomorphic curves in $\mathbb{C}P^n$ is basically the same, just more complicated. This example is considered in Griffiths [G, 1974]. It should be contrasted to the example of holomorphic curves in $\mathbb{C}P^2$ under the action of $PSL(3; \mathbb{C})$, which is considered in Cartan [C, 1937] in the section "Géométrie Projective. Étude des courbe planes," (pp. 204-218).

In this example there are three types of holomorphic curves, called Types: IIa (hyperplane), IIbIIIa (Veronese variety), and IIbIIIb.

We conclude this chapter with a very brief descussion of a geometric interpretation of the invariants and the relationship between these invariants and the first fundamental form of the curve. We state and prove Calabi's theorem which states that the first fundamental form of a holomorphic curve in $\mathbb{C}P^2$ determines the curve up to $SU(3)$-congruence. Since "generic" holomorphic curves in $\mathbb{C}P^2$ require at least fourth order contact for congruence (those of Type IIbIIIbi; two such curves may have third order contact at every point and not be congruent), it is apparent that Calabi's theorem should not be interpreted as meaning that holomorphic curves in $\mathbb{C}P^2$ are determined up to congruence by first order contact.

Now $M = \mathbb{C}P^2$, which we realize as $\mathbb{C}^{3^*}/\mathbb{C}^*$, where $\mathbb{C}^{3^*}$ denotes the set of all non-zero vectors in $\mathbb{C}^3$ and $\mathbb{C}^*$ is the group of non-zero complex numbers. For $z \in \mathbb{C}^{3^*}$ we denote its equivalence class in $\mathbb{C}P^2$ by $[z]$. The group $G = SU(3)$ acts on $\mathbb{C}P^2$ by matrix multiplication from the left. For an origin of M we take $0 = \begin{bmatrix} 1 \\ 0 \\ 0 \end{bmatrix}$, and then the isotropy subgroup of G at 0 is $G_0 = S(U(1) \times U(2))$ $= \left\{ \begin{pmatrix} a & 0 & 0 \\ 0 & & \\ 0 & A & \end{pmatrix} : a \in U(1),\ A \in U(2),\ a \det A = 1 \right\}$. The map $\pi\colon G \to G/G_0 = M$ denotes the natural projection.

The Lie algebra of G is $\mathfrak{g} = su(3)$, and of G_0 is $\mathfrak{g}_0$ $= \left\{ \begin{pmatrix} ix & 0 & 0 \\ 0 & & \\ 0 & Y & \end{pmatrix} : x \in \mathbb{R},\ Y \in u(2),\ ix + \operatorname{Trace} Y = 0 \right\}$. The Maurer-Cartan form of G is $\Omega_A = A^{-1}dA = (\omega^i_j)$, where $\omega^i_j + \bar{\omega}^j_i = 0$, $\omega^i_j = \sum_1^3 A^k_i dA^k_j$, and A^i_j is the ij^{th} entry of $A \in SU(3)$, $1 \leq i$, $j \leq 3$. The Maurer-Cartan equations $d\Omega = -\frac{1}{2}[\Omega, \Omega]$ are $d\omega^i_j = -\sum_1^3 \omega^i_k \wedge \omega^k_j$.

As a vector subspace of $\mathfrak{g}$ complementary to $\mathfrak{g}_0$ we take the $\operatorname{Ad}(G_0)$-invariant subspace $\mathfrak{m}_0 = \left\{ \begin{pmatrix} 0 & -{}^t\bar{z} \\ z & 0 \end{pmatrix} : z \in \mathbb{C}^2 \right\}$, and for a basis of $\mathfrak{m}_0$ over $\mathbb{C}$ we choose $E_1 = \begin{pmatrix} 0 & -1 & 0 \\ 1 & & \\ 0 & & 0 \end{pmatrix}$, $E_2 = \begin{pmatrix} 0 & 0 & -1 \\ 0 & & \\ 1 & & 0 \end{pmatrix}$.

The decomposition $\mathfrak{g} = \mathfrak{g}_0 + \mathfrak{m}_0$ gives $\Omega = \omega_0 + \Theta_0$, where ω_0 is the $\mathfrak{g}_0$-component and $\Theta_0 = \omega_1^2 \otimes E_1 + \omega_1^3 \otimes E_2$. (We note that the action of $\mathbb{C}$ on $\mathfrak{m}_0$ is $a \cdot \begin{pmatrix} 0 & {}^t\!-\bar{z} \\ z & 0 \end{pmatrix} = \begin{pmatrix} 0 & {}^t\!-\overline{az} \\ az & 0 \end{pmatrix}$, where $a \in \mathbb{C}$, $z \in \mathbb{C}^2$. Thus $\Theta_0 = \begin{pmatrix} 0 & \omega_2^1 & \omega_3^1 \\ \omega_1^2 & & \\ \omega_1^3 & & 0 \end{pmatrix}$.)

We consider a holomorphic imbedding $f: S \to M$, where S is a connected, 1-dimensional complex manifold. The zeroth order frames on f are the elements of $L_0 = \{A \in G: \pi(A) \in f(S)\}$, a G_0-bundle over S. A zeroth order frame field along f is a smooth map $u: S \to G$ such that $\pi \circ u = f$. It makes no sense to ask that u be holomorphic since G is not a complex manifold. But the following holds.

<u>Lemma</u>. For any zeroth order frame field u along f, $u^*\Theta_0$ is a bidegree $(1, 0)$ form on S.

<u>Proof</u>: Let X be a type $(0, 1)$ vector at $s \in S$. Then, since f is holomorphic, $0 = f_*X = (\pi \circ u)_*X = \pi_{*_u} u_* X = \pi_{*_u} \cdot u \cdot (u^*\Omega)X$
$= u(s)_{*_0} \pi_{*_1}(u^*\Omega)X = u(s)_{*_0}\pi_{*_1}(u^*\Theta_0)X$, since $\mathfrak{g}_0 = \text{kernel of } \pi_{*_1}$.
Thus $(u^*\Theta_0)X = 0$, since $u(s)_{*_0}\pi_{*_1}: \mathfrak{m}_0 \to M_{u(s)0}$ is an isomorphism.
Hence $u^*\Theta_0$ has bidegree $(1, 0)$.

Let θ be a nowhere vanishing (1, 0) form on S, let u be a zeroth order frame along f and set $u^*\omega_1^2 = x\theta$, $u^*\omega_1^3 = y\theta$, where $x, y: S \to C$ are smooth maps. Any other zeroth order frame field along f is given by $\tilde{u} = uK$, where $K = \begin{pmatrix} a & 0 & 0 \\ 0 & & \\ 0 & A & \end{pmatrix}$, $a: S \to U(1)$, $A: S \to U(2)$ are smooth maps and $a \det A = 1$. Set $\tilde{u}^*\omega_1^2 = \tilde{x}$, $\tilde{u}^*\omega_1^3 = \tilde{y}\theta$, where $\tilde{x}, \tilde{y}: S \to \mathbb{C}$ are smooth maps. Then $\tilde{u}^*\Theta_0 = ad(K^{-1})u^*\Theta_0$ implies that $\begin{pmatrix}\tilde{x}\\ \tilde{y}\end{pmatrix} = A\binom{x}{y}a^{-1}$, or, using brackets to indicate points in $\mathbb{C}P^1$, $\begin{bmatrix}\tilde{x}\\ \tilde{y}\end{bmatrix} = A[{x \atop y}]$. (Since f is an imbedding, neither $\binom{x}{y}$ nor $\begin{pmatrix}\tilde{x}\\ \tilde{y}\end{pmatrix}$ is ever equal to $\binom{0}{0}$.)

The standard action of $U(2)$ on $\mathbb{C}P^1$ is transitive, so the set $W_1 = \{[{1 \atop 0}]\}$ is a cross section. Any holomorphic curve in M is locally the type of W_1; i.e., locally there exists a zeroth order frame field u along f such that $u^*\omega_1^3 = 0$. This condition characterizes the first order frame fields along f. There are no first order invariants.

Geometrically, a first order frame on f at $s \in S$ is a zeroth order frame A at s such that f_*S_s is the complex line in $M_{f(0)}$ spanned over $\mathbb{C}$ by the (1, 0) vector $A_*\pi_*E_1$.

The isotropy subgroup of G_0 at $[{1 \atop 0}]$ is G_1

$= \left\{\begin{pmatrix} a & & \\ & b & \\ & & c\end{pmatrix} : a, b, c \in U(1), \ abc = 1\right\}$. (Blank spaces in matrices will always denote zeros). Its Lie algebra is $\mathfrak{g}_1$

$$= \left\{ \begin{pmatrix} ir & & \\ & is & \\ & & it \end{pmatrix} : r, s, t \in \mathbb{R} , \quad r + s + t = 0 \right\}.$$ For a subspace of $\mathfrak{g}_0$ complementary to $\mathfrak{g}_1$ we take the $Ad(G_1)$-invariant subspace $\mathfrak{m}_1$

$$= \left\{ \begin{pmatrix} 0 & 0 & 0 \\ 0 & 0 & -\bar{z} \\ 0 & z & 0 \end{pmatrix} : z \in \mathbb{C} \right\},$$ and for a basis over $\mathbb{C}$ of it we take E_3

$$= \begin{pmatrix} 0 & 0 & 0 \\ 0 & 0 & -1 \\ 0 & 1 & 0 \end{pmatrix}.$$

The decomposition $\mathfrak{g}_0 = \mathfrak{g}_1 + \mathfrak{m}_1$ decomposes ω_0 into $\omega_0 = \omega_1 + \Theta_1$, where ω_1 is the $\mathfrak{g}_1$-component and $\Theta_1 = \omega_2^3 \otimes E_3$

$$= \begin{pmatrix} 0 & 0 & 0 \\ 0 & 0 & \omega_3^2 \\ 0 & \omega_2^3 & 0 \end{pmatrix}.$$

Let u be a first order frame field along f. Then $u^*\omega_1^3 = 0$, and we set $\theta = u^*\omega_1^2$, a nowhere vanishing bidegree $(1, 0)$ form on S. Taking the exterior derivative of $u^*\omega_1^3 = 0$ and applying the Maurer-Cartan equations, it follows that $u^*\omega_2^3$ has bidegree $(1, 0)$. Thus we set $u^*\omega_2^3 = k\theta$, where k is a complex valued smooth function on S.

Any other first order frame field $\tilde{u}$ along f is given by $\tilde{u}$

$$= uK, \quad \text{where } K = \begin{pmatrix} a & & \\ & b & \\ & & c \end{pmatrix},$$ $a, b, c\colon S \to U(1)$ smooth maps such that $abc = 1$. Let $\tilde{\theta} = \tilde{u}^*\omega_1^2$, and $\tilde{u}^*\omega_2^3 = \tilde{k}\,\tilde{\theta}$. Then $\tilde{u}^*(\Theta_0 + \Theta_1) = ad(K^{-1})u^*(\Theta_0 + \Theta_1)$ implies that $\tilde{\theta} = \bar{a}b\,\theta$ and $\tilde{k} = \bar{b}^3 k$. This

action of $U(1)$ on $\mathbb{C}$ given by $z \mapsto b^3 z$ has two orbit types: the origin, (Type IIa), and the complement of the origin, (Type IIb).

Type IIa curves.

We say $f: S \to M$ is Type IIa if for some, hence any, first order frame field u along f $u^*\omega_2^3 = 0$. Then there are no second order invariants, $G_2 = G_1$, and $L_j = L_1$ for all $j \geq 1$. By the Corollary of section I.12, $f(S)$ is G-congruent to an open submanifold of $\pi(H)$, where H is the analytic subgroup of G whose Lie algebra is $\mathfrak{h}$

$$= \{\omega_1^3 = 0\ ,\ \omega_2^3 = 0\} = \left\{\begin{pmatrix} X & & 0 \\ & & 0 \\ 0 & 0 & x \end{pmatrix} : X \in u(2),\ x \in u(1),\ x + \text{Trace } X = 0\right\}$$

$= s(u(2) \times u(1))$. Thus $H = S(U(2) \times U(1))$ and $\pi(H)$

$$= \left\{\begin{pmatrix} A & & 0 \\ & & 0 \\ 0 & 0 & a \end{pmatrix}\begin{bmatrix} 1 \\ 0 \\ 0 \end{bmatrix} : A \in U(2),\ a \in U(1),\ a \det A = 1\right\} = \left\{\begin{bmatrix} z \\ \\ 0 \end{bmatrix} : z \in \mathbb{C}^{2*}\right\},$$

which is a hyperplane in M.

Type IIb curves.

We say $f: S \to M$ is Type IIb if for some, hence any, first order frame field u along f $u^*\omega_2^3$ never vanishes. Then locally there always exists a first order frame field u such that $k > 0$. This characterizes the second order frames, and $k: S \to \mathbb{R}^+$ is the second order invariant.

The isotropy subgroup of G_1 at any point of the positive real axis is $G_2 = \left\{ \begin{pmatrix} a & & \\ & b & \\ & & c \end{pmatrix} \in G_1 : \bar{b}^3 = 1 \right\}$

$= \left\{ \begin{pmatrix} a & & \\ & \varepsilon & \\ & & \overline{a\varepsilon} \end{pmatrix} : \varepsilon,\ a \in U(1),\ \varepsilon^3 = 1 \right\}$. G_2 has three connected components, corresponding to three orientation classes of second order frames. The Lie algebra of G_2 is $\mathfrak{g}_2 = \left\{ \begin{pmatrix} i & & \\ & 0 & \\ & & -i \end{pmatrix} t : t \in \mathbb{R} \right\}$. For a subspace of $\mathfrak{g}_1$ complementary to $\mathfrak{g}_2$ we take the $\mathrm{ad}(G_2)$-invariant subspace $\mathfrak{m}_2 = \left\{ \begin{pmatrix} it & 0 & 0 \\ 0 & -2it & 0 \\ 0 & 0 & it \end{pmatrix} : t \in \mathbb{R} \right\}$. Then the $\mathfrak{g}_1$-component of Ω, ω_1, decomposes into $\omega_1 = \omega_2 + \Theta_2$, where $\Theta_2 = \begin{pmatrix} -\frac{1}{2}\omega_2^2 & 0 & 0 \\ 0 & \omega_2^2 & 0 \\ 0 & 0 & -\frac{1}{2}\omega_2^2 \end{pmatrix}$.

Let u be a second order frame field along f and set $u^*\omega_1^2 = \theta$, $u^*\omega_2^3 = k\theta$, $k: S \to \mathbb{R}^+$, $u^*\omega_2^2 = r\theta - \overline{r\theta}$, and $dk = k_{;1}\theta + \overline{k_{;1}}\,\bar{\theta}$, where $r, k_{;1}: S \to \mathbb{C}$ are smooth functions. If we take the exterior derivative of both sides of $u^*\omega_2^3 = k\theta$ and apply the Maurer-Cartan equations $d\omega_2^3 = -\sum_j \omega_j^3 \wedge \omega_2^j$, it follows that $k_{;1} = -3kr$.

Let $\tilde{u}$ be any other second order frame field and define $\tilde{\theta}$, $\tilde{r}$, and $\tilde{k}_{;1}$ as for u. Then $\tilde{u} = uK$, where $K = \begin{pmatrix} a & 0 & 0 \\ 0 & \varepsilon & 0 \\ 0 & 0 & \overline{a\varepsilon} \end{pmatrix}$, $a: S \to U(1)$ is a smooth map, and ε is a cube root of 1. From $\tilde{u}^*(\Theta_0 + \Theta_1 + \Theta_2) = \mathrm{ad}(K^{-1})u^*(\Theta_0 + \Theta_1 + \Theta_2)$ it follows that $\tilde{\theta} = \varepsilon\bar{a}\,\theta$,

$\tilde{r}\,\tilde{\theta} - r\,\overline{\tilde{\theta}} = r\,\theta - \overline{r\,\theta}$, $\tilde{r} = \bar{\varepsilon}ar$, and $\tilde{k}_{;1} = \bar{\varepsilon}ak_{;1}$. In particular the form $u^*\omega_2^2 = r\,\theta - \overline{r\,\theta}$ on S is independent of the choice of second order frame field along f .

The action of $U(1) \times \{\varepsilon \in U(1): \varepsilon^3 = 1\}$ on $\mathbb{C}$ given by $(a, \varepsilon)z = az\bar{\varepsilon}$ has two orbit types: the origin and its complement.

Type IIbIIIa.

We say that $f: S \to M$ is of Type IIbIIIa if it is of Type IIb and for some, hence any, second order frame field u along f we have $u^*\omega_2^2 = 0$. Consequently $r = 0$ and $k_{;1} = -3kr = 0$, so k is a positive constant. There are no third order invariants. Taking the exterior derivative of $u^*\omega_2^2 = 0$ and applying the Maurer-Carton equations, it follows that $k = 1$.

There are no non-constant invariants, so by the Corollary of section I.12 f(S) is G-congruent to an open submanifold of $\pi(H)$, where H is the analytic subgroup of G whose Lie algebra is $\mathfrak{h}$ $= \{\omega_1^3 = 0 , \omega_2^3 = \omega_1^2 , \omega_2^2 = 0\}$. The homogeneous curve $\pi(H)$ is just the Veronese variety $g(\mathbb{C}P^1)$, where $g: \mathbb{C}P^1 \to \mathbb{C}P^2$ is the Veronese map $g[{z \atop w}] = \begin{bmatrix} z^2 \\ \sqrt{2}\, zw \\ w^2 \end{bmatrix}$. We demonstrate this as follows.

In order to identify the subgroup H , consider the representation of SU(2) on the space $\mathcal{S}_2(\mathbb{C}^2)$, the symmetric product of $\mathbb{C}^2$, induced from the standard representation of SU(2) on $\mathbb{C}^2$. Namely, if $x, y \in \mathbb{C}^2$, $A \in SU(2)$ and xy denotes the symmetric product of x and y, then $A(xy) = AxAy$. Let $\rho: SU(2) \to GL(3; \mathbb{C})$ be the matrix representation obtained using the basis ε_1^2, $\sqrt{2}\,\varepsilon_1\varepsilon_2$, ε_2^2 of $\mathcal{S}_2(\mathbb{C}^2)$, where $\varepsilon_1, \varepsilon_2$ is the standard basis of $\mathbb{C}^2$. A straightforward computation shows that $\rho_* su(2) = \mathfrak{h}$, and thus $\rho(SU(2)) = H$. Furthermore, for any $A \in SU(2)$, $g \circ A = \rho(A) \circ g$, where $g: \mathbb{C}P^1 \to \mathbb{C}P^2$ is the Veronese map. Hence $\pi(H) = H \cdot \begin{bmatrix}1\\0\\0\end{bmatrix}$ $= \rho(SU(2)) \cdot g\left[\begin{smallmatrix}1\\0\end{smallmatrix}\right] = g(SU(2) \cdot \left[\begin{smallmatrix}1\\0\end{smallmatrix}\right]) = g(\mathbb{C}P^1)$.

Type IIbIIIb.

We say that the Type IIb curve $f: S \to M$ is of Type IIbIIIb if for some, hence any, second order frame field u along f the form $u^*\omega_2^2 \neq 0$; i.e., $r \neq 0$. A cross section for the above action of $U(1) \times \{\varepsilon \in U(1): \varepsilon^3 = 1\}$ on $\mathbb{C}^*$ is $W_3 = \{z \in \mathbb{C}: z > 0\}$. There exists, at least locally, a second order frame field u along f such that $r > 0$, where $u^*\omega_2^2 = r\theta - \overline{r\,\theta}$ as defined above. This characterizes the third order frame fields along f. The smooth function $r: S \to \mathbb{R}^+$ is a third order invariant. The other third order invariant is $k_{;1} = -3kr < 0$.

The isotropy subgroup of G_2 at any point of W_3 is G_3

$$= \left\{ \begin{pmatrix} a & & \\ & \varepsilon & \\ & & a\bar{\varepsilon} \end{pmatrix} \in G_2 : \bar{\varepsilon} a > 0 \right\} = \{\varepsilon I_3 : \varepsilon \in U(1),\ \varepsilon^3 = 1\},$$

a three element group. The third order frames are uniquely determined up to three orientation classes.

A complementary subspace to $\mathfrak{g}_3$ in $\mathfrak{g}_2$ is $\mathfrak{m}_3 = \mathfrak{g}_2$, and

thus $\omega_2 = \theta_3 = \begin{bmatrix} \omega_1^1 + \frac{1}{2}\omega_2^2 & 0 & 0 \\ 0 & 0 & 0 \\ 0 & 0 & \omega_3^3 + \frac{1}{2}\omega_2^2 \end{bmatrix}$. If u is any third order

frame field along f, we set $u^*\omega_1^1 = z\theta - \bar{z}\bar{\theta}$, where the smooth function $z \colon S \to \mathbb{C}$ is a fourth order invariant.

The adjoint action of G_3 on $\mathfrak{m}_0 + \mathfrak{m}_1 + \mathfrak{m}_2 + \mathfrak{m}_3 = \mathfrak{g}$ is trivial, so all invariants are "orientation invariant" and $L_j = L_3$ for all $j \geqslant 3$. If u is a third order frame field along f, then $\theta = u^*\omega_1^2$ is a bidegree $(1, 0)$ form on S, $u^*\omega_2^3 = k\theta$ defines the non-constant, positive second order invariant k, $u^*\omega_2^2 = r(\theta - \bar{\theta})$ defines the positive third order invariant r, and $u^*\omega_1^1 = z\theta - \bar{z}\bar{\theta}$ defines the complex valued fourth order invariant z.

The structure equations of f are:

$$d\theta = (\bar{z} - r)\theta \wedge \bar{\theta}, \qquad k_{;1} = -3kr,$$

$$\operatorname{Re} r_{;1} = \frac{1}{2}(k^2 - 1) + r \operatorname{Re} z - r^2,$$

$$\operatorname{Re} z_{;2} = \frac{1}{2} + |z|^2 - r \operatorname{Re} z,$$

where as usual $dh = h_{;1}\theta + h_{;2}\bar{\theta}$ for any smooth function h on S. We denote real part by Re and imaginary part by Im.

Further relations are imposed by the equations $ddk = 0$ and $ddr = 0$. These are: $\mathrm{Im}\, r_{;1} = r\, \mathrm{Im}\, z$ and $r\, \mathrm{Im}\, z_{;2} = (1 - k^2 - r^2)\, \mathrm{Im}\, z$. Thus the third structure equation holds even when "Re" is omitted from both sides.

The fine structure of Type IIbIIIb has two parts. Observe that $k > 0$ is non-constant for this type. It is easily checked that $dk \wedge dr = 0$ on S iff $\mathrm{Im}\, r_{;1} = 0$, which in turn occurs iff $\mathrm{Im}\, z = 0$. It follows that either (i) k and r are independent, or (ii) all invariants are (locally) functions of k.

Type IIbIIIb(i).

These are Type IIbIIIb curves for which k and r are independent, the "generic" curves. The Frenet frames have order 3; that is, at least fourth order contact at every point is required for congruence.

Type IIbIIIb(ii).

These are Type IIbIIIb curves for which all invariants are functions of k. The Frenet frames have order three. From the remarks above it follows that all invariants are real. If $r = r(k)$, $z = z(k)$, then $r_{;1} = -3krr'$ and $z_{;1} = -3krz'$, etc., where primes denote differentiation with respect to k. Then the third structure equation expresses z as a function of r and k, and substituting into the fourth structure equations gives a non-linear second order differential equation which must be satisfied by $r(k)$.

By the Homogeneity Theorem we know that $f(S)$ has a 1-parameter group of local transformations H acting on it. We can identify H as the analytic subgroup of G whose Lie algebra is $\mathfrak{h}$, where $\mathfrak{h}$ is defined as follows. Translating f by an element of G, if necessary, we may fix a point $s_0 \in S$ such that $f(s_0) = 0$ and $u(s_0) = 1$ for some third order frame field u along f. Setting $k_0 = k(s_0)$, $r_0 = r(s_0)$, $z_0 = z(s_0)$, we have $\mathfrak{h}$

$$= \{\omega_1^3 = 0\ ,\ \omega_2^3 = k_0\omega_1^2\ ,\ \omega_1^2 + \bar{\omega}_1^2 = 0\ ,\ \omega_2^2 = 2r_0\omega_1^2\ ,\ \omega_1^1 = 2z_0\omega_1^2\}\ .$$

These are the W-curves of Klein and Lie, (cf. Klein [K, 1926], pp. 126-128 and 166-173). In fact, $\mathfrak{h}$

$$= \left\{ \begin{pmatrix} 2z_0 & 1 & 0 \\ 1 & 2r_0 & k_0 \\ 0 & k_0 & -2(z_0+r_0) \end{pmatrix} it\colon t \in \mathbb{R} \right\} .$$ The matrix $X =$

$$= \begin{pmatrix} 2z_0 & 1 & 0 \\ 1 & 2r_0 & k_0 \\ 0 & k_0 & -2(z_0+r_0) \end{pmatrix}$$ is real and symmetric, so there exists

$A \in SO(3)$ such that $AX{}^tA = \begin{pmatrix} a & 0 & 0 \\ 0 & b & 0 \\ 0 & 0 & c \end{pmatrix}$, where $a, b, c \in \mathbb{R}$, $a + b + c = 0$. If H is the analytic subgroup of G whose Lie algebra is $\mathfrak{h}$, then $H = A^{-1}H'A$, where $H' = \left\{ \begin{pmatrix} e^{ita} & 0 & 0 \\ 0 & e^{itb} & 0 \\ 0 & 0 & e^{itc} \end{pmatrix} : t \in \mathbb{R} \right\}$

Then $f(S)$ is (locally) invariant under H implies that $Af(S)$ is (locally) invariant under H' .

Locally $A \circ f(S)$ is defined by $F(x_1, x_2, x_3) = \text{const.}$, for some holomorphic function F , (e.g., lift $A \circ f$ to $\mathbb{C}^3$ and use the implicit function theorem). Then F is invariant under the action of H' , so $0 = \frac{d}{dt} F(e^{ita}x_1, e^{itb}x_2, e^{itc}x_1) = i(ay_1F_1 + by_2F_2 + cy_3F_3)$, where $y_1 = e^{ita}x_1$, $y_2 = e^{itb}x_2$, $y_3 = e^{itc}x_3$, and $F_j = \frac{\partial F}{\partial y_j}$. It follows that $F(y_1, y_2, y_3) = y_1^{b-c}y_2^{c-a}y_3^{a-b}$.

Geometric interpretation of k .

If S is a complex 1-manifold with hermitian metric g , which is given locally by $g = \theta\,\overline{\theta}$, where θ is a nowhere vanishing bidegree $(1, 0)$ form on S , then the Levi-Civita connection form ϕ of g is the unique $u(1)$-valued 1-form satisfying the equation $d\theta = -\phi \wedge \theta$. The Gaussian curvature K satisfies the equation $d\phi = \frac{1}{2} K\theta \wedge \overline{\theta}$.

The Fubini-Study metric on $\mathbb{C}P^2$ is the G-invariant hermitian metric defined by the $Ad(G_0)$-invariant hermitian form $\omega_1^2\overline{\omega}_1^2 + \omega_1^3\overline{\omega}_1^3$ on G . Explicitly, if u is a local adapted frame field on M , i.e., a local cross section of $\pi: G \to M$, then the Fubini-Study metric is $ds^2 = u^*(\omega_1^2\overline{\omega}_1^2 + \omega_1^3\overline{\omega}_1^3)$.

Let $f: S \to M$ be a holomorphic curve, and let u be a first order frame field along f . Then $\theta = u^*\omega_1^2$ is a non-vanishing $(1, 0)$ form on S , $u^*\omega_1^3 = 0$, and $u^*\omega_2^3 = k\theta$ for some smooth function

$k: S \to \mathbb{C}$. Then the induced hermitian metric on S is $f^*ds^2 = \theta\bar{\theta}$. Differentiating $u^*\omega_1^2 = \theta$ and applying the Maurer-Cartan equations, it follows that the Levi-Civita connection form is $\phi = u^*(\omega_2^2 - \omega_1^1)$. Then $d\phi = (2 - |k|^2)\theta \wedge \bar{\theta}$ implies that the Gaussian curvature $K = 2(2 - |k|^2)$.

The following theorem is due to Calabi and is true for holomorphic curves in $\mathbb{C}P^n$, for all n. It is stated and proved in Griffiths [G, 1974] p. 792.

<u>Theorem</u>. Let $f, \tilde{f}: S \to M$ be holomorphic curves in M and let ds^2 denote the Fubini-Study metric on M. If $f^*ds^2 = \tilde{f}^*ds^2$, then f and $\tilde{f}$ are G-congruent.

<u>Remark</u>: The statement of this theorem is not equivalent to (5.3) on p. 800 of Griffiths [G, 1974]. In fact, since there is no first order invariant, any two holomorphic curves in M have at least first order G-contact between any two points. As we've seen above, for a Type IIbIIIb curve at least fourth order contact is required to imply congruence.

<u>Proof of Theorem</u>: Tilde'd quantities always refer to $\tilde{f}$. Now $f^*ds^2 = \tilde{f}^*ds^2$ implies that $2(2 - k^2) = K = \tilde{K} = 2(2 - \tilde{k}^2)$, and hence $k = \tilde{k}$. Thus the theorem is proved for Type IIa and Type IIbIIIa curves.

Suppose now that f and $\tilde{f}$ are of Type IIbIIIb. Let u and $\tilde{u}$ be third order frame fields along f and $\tilde{f}$, respectively, and use the notation preceding the structure equations above for Type IIbIIIb curves. Then $\theta\bar{\theta} = \tilde{\theta}\bar{\tilde{\theta}}$ implies there exists a smooth map $a: S \to U(1)$ such that $\bar{a}\tilde{\theta} = \theta$ on S. Define a second order frame $\tilde{v}$ along $\tilde{f}$ by $\tilde{v} = \tilde{u}A$, where

$$A = \begin{pmatrix} a & 0 & 0 \\ 0 & 1 & 0 \\ 0 & 0 & \bar{a} \end{pmatrix} \in G_2 ,$$

Then $\tilde{v}^*\omega_1^2 = \bar{a}\tilde{\theta} = \theta$ and $\tilde{v}^*\omega_2^3 = \tilde{k}\theta = k\theta$, since $\tilde{k}$ is a second order invariant. Set $\tilde{v}^*\omega_2^2 = x\theta - \bar{x}\bar{\theta}$, where $x: S \to \mathbb{C}$ is a smooth function. Taking the exterior derivative of $u^*\omega_2^3 = k\theta$ and of $\tilde{v}^*\omega_2^3 = k\theta$, and applying the Maurer-Cartan equations, it follows that $k_{;1} = -3kr$ and $k_{;1} = -3kx$. Hence $x = r > 0$ and it follows that $\tilde{v}$ must in fact be a third order frame field along $\tilde{f}$. Hence A must actually take values in G_3, so $A = I$ and $\tilde{v} = \tilde{u}$, $\tilde{\theta} = \theta$ and $\tilde{r} = x = r$. Finally, from the first structure equation, $d\theta = (\bar{y} - \frac{3}{2} r)\theta \wedge \bar{\theta} = (\bar{\tilde{y}} - \frac{3}{2} r)\theta \wedge \bar{\theta}$, which implies that $\tilde{y} = y$. By Corollary 1 of section I.11, f and f are G-congruent.

V. Holomorphic curves in complex Grassmannians.

As in the previous examples, we simplify the exposition by considering the case of holomorphic curves in $\mathbb{C}G_{4,2}$, the complex Grassmann manifold of complex 2-planes in $\mathbb{C}^4$, under the action of $SU(4)$. The analysis of curves in $\mathbb{C}G_{n,q}$, for $1 \leq q \leq [\frac{n}{2}]$, is similar, but more complicated. This example is considered in Griffiths [G, 1974].

Now $M = \mathbb{C}G_{4,2}$, which we realize as $\mathbb{C}^{4\times 2^*}/GL(2;\mathbb{C})$, where $\mathbb{C}^{4\times 2}$ is the space of 4×2 complex matrices, $\mathbb{C}^{4\times 2^*}$ is the open subset of those of rank 2, and $GL(2;\mathbb{C})$ acts on $\mathbb{C}^{4\times 2^*}$ on the right by matrix multiplication. If $Z \in \mathbb{C}^{4\times 2^*}$, we let $[Z]$ denote its equivalence class in $\mathbb{C}G_{4,2}$.

The group $G = SU(4)$ acts on M by matrix multiplication from the left. For the origin of M we take $o = \begin{bmatrix} 1 & 0 \\ 0 & 1 \\ 0 & 0 \\ 0 & 0 \end{bmatrix}$, and then the isotropy subgroup of G at o is $G_0 = S(U(2) \times U(2))$.

The Lie algebra of G is $\mathfrak{g} = su(4)$, and of G_0 is $\mathfrak{g}_0 = s(u(2) \times u(2))$. The Maurer-Cartan form of G is $\Omega_A = A^{-1}dA = (\omega^i_j)$, where $\omega^i_j + \bar{\omega}^j_i = 0$, $\omega^i_j = \sum_1^4 \bar{A}^k_i dA^k_j$, A^i_j is the $ij^{\underline{th}}$ entry of $A \in SU(4)$, and $1 \leq i, j \leq 4$. The Maurer-Cartan equations $d\Omega = -\frac{1}{2}[\Omega, \Omega]$ are $d\omega^i_j = -\sum_1^4 \omega^i_k \wedge \omega^k_j$.

As a vector subspace of $\mathfrak{g}$ complementary to $\mathfrak{g}_0$ we take the $\mathrm{Ad}(G_0)$-invariant subspace $\mathfrak{m}_0 = \left\{ \begin{pmatrix} 0 & {}^t\bar{z} \\ z & 0 \end{pmatrix} : z \in \mathbb{C}^{2\times 2} \right\}$. The decomposition $\mathfrak{g} = \mathfrak{g}_0 + \mathfrak{m}_0$ gives $\Omega = \omega_0 + \Theta_0$, where ω_0 is the $\mathfrak{g}_0$-component and

$$\Theta_0 = \begin{pmatrix} 0 & & \omega_3^1 & \omega_4^1 \\ & & \omega_3^2 & \omega_4^2 \\ \omega_1^3 & \omega_2^3 & & \\ \omega_1^4 & \omega_2^4 & & 0 \end{pmatrix}$$

is the $\mathfrak{m}_0$-component.

We consider a holomorphic imbedding $f: S \to M$, where S is a connected, 1-dimensional complex manifold. The zeroth order frames on f are the elements of $L_0 = \{A \in G: \pi(A) \in f(S)\}$, where $\pi: G \to G/G_0 = M$ is the natural projection. A zeroth order frame field along f is a real analytic map $u: S \to G$ such that $\pi \circ u = f$.

As in Example IV, the holomorphicity of f is reflected in zeroth order frame fields along f in the following way.

<u>Lemma 1</u>. For any zeroth order frame field $u: S \to G$ along f the $\mathfrak{m}_0$-valued 1-form $u^*\Theta_0$ on S has bidegree $(1, 0)$.

Let θ be a nowhere vanishing $(1, 0)$ form on S, let u be a zeroth order frame field along f and set $u^*\omega_j^{2+i} = x_j^i\theta$, where x_j^i are complex-valued smooth functions on S, $1 \leq i$, $j \leq 2$. Let $X: S \to \mathbb{C}^{2\times 2}$ be the matrix-valued function whose entries are the functions x^i_j . Any other zeroth order frame field along f is given by $\tilde{u} = uK$, where $K = \begin{pmatrix} A & 0 \\ 0 & B \end{pmatrix}$, $A, B: S \to U(2)$ are smooth maps and $\det A \det B = 1$. Set $\tilde{u}^*\omega^{2+i}_j = \tilde{x}^i_j\theta$ for $1 \leq i$, $j \leq 2$, and let

$$\tilde{X} = \begin{bmatrix} \tilde{x}^1_1 & \tilde{x}^1_2 \\ \tilde{x}^2_1 & \tilde{x}^2_2 \end{bmatrix} .$$ Then $\tilde{u}^*\theta_0 = \mathrm{ad}(K^{-1})u^*\theta_0$ implies that $\tilde{X} = B^{-1}XA$.

We represent $\mathbb{C}P^3$ by $\mathbb{C}^{2\times 2*}$ modulo scalar multiplication by non-zero complex numbers, and denote elements of $\mathbb{C}P^3$ by $[Y]$, where $Y \in \mathbb{C}^{2\times 2*}$.

<u>Lemma 2</u>: $W_1 = \{[\begin{smallmatrix} 1 & 0 \\ 0 & k \end{smallmatrix}]: 0 \leq k \leq 1\}$ is a cross section of the action of G_0 on $\mathbb{C}P^3$ given by $(A, B)[X] = [BXA^{-1}]$, where $X \in \mathbb{C}^{2\times 2*}$, $A, B \in U(2)$, $\det A \det B = 1$.

The action of G_0 on $\mathbb{C}P^3$ given in Lemma 2 has three orbit types: those meeting $\{[\begin{smallmatrix} 1 & 0 \\ 0 & 0 \end{smallmatrix}]\}$ (Type Ia), those meeting $\{[\begin{smallmatrix} 1 & 0 \\ 0 & 1 \end{smallmatrix}]\}$ (Type Ib), and those meeting $W_1 = \{[\begin{smallmatrix} 1 & 0 \\ 0 & k \end{smallmatrix}]: 0 < k < 1\}$ (Type Ic).

<u>Type Ia</u>.

We say $f: S \to M$ is of Type Ia if there is a zeroth order frame field u along f for which $u^*\omega^3_2 = u^*\omega^4_1 = u^*\omega^4_2 = 0$. This condition characterizes the first order frame fields along f . There are no first order invariants.

The isotropy subgroup of G_0 at $\begin{bmatrix}1 & 0\\ 0 & 0\end{bmatrix}$ is

$$G_1 = \left\{\begin{pmatrix}A & 0\\ 0 & B\end{pmatrix} \in G_0 : B^{-1}\begin{bmatrix}1 & 0\\ 0 & 0\end{bmatrix}A = \begin{bmatrix}1 & 0\\ 0 & 0\end{bmatrix}\right\}$$

$$= \left\{\begin{pmatrix}a & & & \\ & b & & \\ & & c & \\ & & & e\end{pmatrix} : a, b, c, e \in U(1),\ abce = 1\right\}.$$ Its Lie algebra is

$$\mathfrak{g}_1 = \left\{\begin{pmatrix}it & & & \\ & ir & & \\ & & is & \\ & & & iq\end{pmatrix} : t, r, s, q \in \mathbb{R},\ t + r + s + q = 0\right\}.$$ For a subspace of $\mathfrak{g}_0$ complementary to $\mathfrak{g}_1$ we take the $\mathrm{Ad}(G_1)$-invariant subspace $$\mathfrak{m}_1 = \left\{\begin{pmatrix}0 & -\bar{z} & & \\ z & 0 & & 0 \\ & & 0 & -\bar{w}\\ & 0 & w & 0\end{pmatrix} : z, w \in \mathbb{C}\right\}.$$ The decomposition $\mathfrak{g}_0 = \mathfrak{g}_1 + \mathfrak{m}_1$ decomposes ω_0 into $\omega_0 = \omega_1 + \Theta_1$, where ω_1 is the $\mathfrak{g}_1$-component and $$\Theta_1 = \begin{pmatrix}0 & \omega_2^1 & & \\ \omega_1^2 & 0 & & 0 \\ & & 0 & \omega_4^3\\ & 0 & \omega_3^4 & 0\end{pmatrix}$$ is the $\mathfrak{m}_1$-component.

If u is a first order frame field along f, we set $u^*\omega_1^3 = \theta$, a non-vanishing bidegree $(1, 0)$ form on S. Taking the exterior derivative of the equations $u^*\omega_2^3 = 0$, $u^*\omega_1^4 = 0$ and $u^*\omega_2^4 = 0$, and applying the Maurer-Cartan equations of G, it follows that $u^*\omega_2^1$ and $u^*\omega_3^4$ are bidegree $(1, 0)$ forms on S. Thus we set $u^*\omega_2^1 = y\theta$

and $u^*\omega_3^4 = z\theta$, where y and z are complex valued smooth functions on S .

Any other first order frame field $\tilde{u}$ along f is given by $\tilde{u} = uK$, where $K = \begin{pmatrix} a & & & \\ & b & & \\ & & c & \\ & & & e \end{pmatrix}$, $a, b, c, e: S \to U(1)$ are smooth maps with $abce = 1$. Let $\tilde{\theta}$, $\tilde{y}$, $\tilde{z}$ be the analogs of θ, y , z for $\tilde{u}$. Then $\tilde{u}^*(\Theta_0 + \Theta_1) = ad(K^{-1})u^*(\Theta_0 + \Theta_1)$ implies that $\tilde{\theta} = a\bar{c}\theta$, $\tilde{y} = \bar{a}^{3}\bar{e}y$, and $\tilde{z} = bc^3z$.

The action of G_1 on $\mathbb{C}^2$ given by $(a, b, c, e)(y, z) = (a^3ey, \overline{bc}^3z)$ has three orbit types: the origin: $y = 0 = z$ (Type IaIIa); a coordinate line (off the origin): y or $z = 0$, but not both (Type IaIIa); and the complement of the coordinate lines: $yz \neq 0$ (Type IaIIc).

Type IaIIa

We say a Type Ia curve $f: S \to M$ is of Type IaIIa if for some, hence any, first order frame field u along f we have $u^*\omega_2^1 = 0 = u^*\omega_3^4$, i.e., $y = 0 = z$. This characterizes the second order frame fields along f , and conseuqently $L_2 = L_1$.

The isotropy subgroup of G_1 at $(0, 0)$ is $G_2 = G_1$. There are no invariants and $L_j = L_1$ for all $j \geq 1$. By the Corollary of section I.12, f is G-congruent to an open submanifold of the homogeneous submanifold $\pi(H)$, where H is the analytic subgroup of G whose Lie algebra is $\mathfrak{h} = \{\omega_2^3 = 0 , \omega_1^4 = 0 , \omega_2^4 = 0 , \omega_2^1 = 0 , \omega_3^4 = 0\}$.

It is easy to see that $\mathfrak{h}$ is the Lie algebra direct sum

$$su(2) \oplus \left\{ \begin{pmatrix} it & & & \\ & is & & \\ & & it & \\ & & & -i(2t+s) \end{pmatrix} : t, s \in \mathbb{R} \right\}, \text{ where } su(2) \text{ is im-}$$

bedded in $su(4)$ by $\rho_* \begin{pmatrix} it & -\bar{w} \\ w & -it \end{pmatrix} = \begin{pmatrix} it & 0 & -\bar{w} & 0 \\ 0 & 0 & 0 & 0 \\ w & 0 & -it & 0 \\ 0 & 0 & 0 & 0 \end{pmatrix}$, $t \in \mathbb{R}$,

$w \in \mathbb{C}$. Thus $H = \rho(SU(2)) \times A$, where A is an abelian subgroup of G_0, and therefore $\pi(H) = \pi\rho(SU(2))$.

Let $h: \mathbb{C}P^1 \to \mathbb{C}G_{4,2}$ be defined by $h\begin{bmatrix} z \\ w \end{bmatrix} = \begin{bmatrix} z & 0 \\ 0 & 1 \\ w & 0 \\ 0 & 0 \end{bmatrix}$, a holomorphic curve of Type IaIIa. If $a \in SU(2)$, $p \in \mathbb{C}P^1$, then $h(a\cdot p) = \rho(a)\cdot h(p)$

Hence $\pi(H) = \pi\rho(SU(2)) = \rho(SU(2))\cdot \begin{bmatrix} 1 & 0 \\ 0 & 1 \\ 0 & 0 \\ 0 & 0 \end{bmatrix} = \rho(SU(2))\cdot h\begin{bmatrix} 1 \\ 0 \end{bmatrix} = h(SU(2)\cdot \begin{bmatrix} 1 \\ 0 \end{bmatrix})$

$= h\mathbb{C}P^1$.

We have shown that any Type IaIIa curve in M is G-congruent to an open submanifold of $h(\mathbb{C}P^1)$.

Type IaIIb.

We say a Type Ia curve $f: S \to M$ is of Type IaIIb if for some, hence any, first order frame field u along f we have $u^*\omega_2^1 = 0$ but $u^*\omega_3^4$ never vanishes (or $u^*\omega_2^1 \neq 0$, $u^*\omega_3^4 = 0$); i.e., $y = 0$, $z \neq 0$ (or $y \neq 0$, $z = 0$). To be specific we assume the former case holds. A cross section for the action of G_1 is then $W_2 = \{(0, z): z > 0\}$. There exists (at least locally) a first order frame field u along f such that $z > 0$. This condition characterizes the second order frame fields along f. The smooth function $z: S \to \mathbb{R}^+$ is the second order invariant.

The isotropy subgroup of G_1 at any point of W_2 is G_2

$$= \left\{\begin{pmatrix} a & & & \\ & b & & \\ & & c & \\ & & & e \end{pmatrix} \in G_1 : bc^3 = 1\right\} = \left\{\begin{pmatrix} a & & & \\ & \bar{c}^3 & & \\ & & c & \\ & & & \bar{a}c^2 \end{pmatrix} : a, c \in U(1)\right\}.$$

Its Lie algebra is $\mathfrak{g}_2 = \left\{\begin{pmatrix} it & & & \\ & -3is & & \\ & & is & \\ & & & i(2s-t) \end{pmatrix} : s, t \in \mathbb{R}\right\}.$

For a subspace of $\mathfrak{g}_1$ complementary to $\mathfrak{g}_2$ we take the $\mathrm{Ad}(G_2)$-invariant subspace $\mathfrak{m}_2 = \left\{\begin{pmatrix} i & & & \\ & 0 & & \\ & & -2i & \\ & & & i \end{pmatrix} r : r \in \mathbb{R}\right\}$. The decomposition $\mathfrak{g}_1 = \mathfrak{g}_2 + \mathfrak{m}_2$ decomposes ω_1 into $\omega_1 = \omega_2 + \Theta_2$, where ω_2 is $\mathfrak{g}_2$-component and $\Theta_2 = \frac{\zeta}{6}\begin{pmatrix} -1 & & & \\ & 0 & & \\ & & 2 & \\ & & & -1 \end{pmatrix}$, $\zeta = \omega_2^2 + 3\omega_3^3$.

Let u be a second order frame field along f, let $u^*\omega_1^3 = \theta$, $u^*\omega_3^4 = z\theta$, $z\colon S \to \mathbb{R}^+$; and set $u^*(\omega_2^2 + 3\omega_3^3) = s\theta - \bar{s}\bar{\theta}$, where $s\colon S \to \mathbb{C}$. Taking the exterior derivative of $u^*\omega_3^4 = z\theta$, and applying the Maurer-Cartan equations of G, it follows that $z_{;1} = sz$. (Doing the same to $u^*\omega_2^1 = 0$ yields no relations.)

Let $\tilde{u} = uK$ be any other second order frame field along f,

$$\text{where } K = \begin{pmatrix} a & & & \\ & \bar{c}^3 & & \\ & & c & \\ & & & \bar{a}c^2 \end{pmatrix}, \quad a, c\colon S \to U(1) \text{ are smooth maps. Let}$$

$\tilde{\theta}$, $\tilde{z}$, $\tilde{s}$ be the analogs of θ, z, s for $\tilde{u}$. Then $\tilde{u}^*(\Theta_0 + \Theta_1 + \Theta_2) = \mathrm{ad}(K^{-1})u^*(\Theta_0 + \Theta_1 + \Theta_2)$ implies that $\tilde{\theta} = a\bar{c}\theta$, and $\tilde{s} = \bar{a}cs$.

The action of G_2 on $\mathbb{C}$ given by $(a, c)s = a\bar{c}s$, $a, c \in U(1)$, $s \in \mathbb{C}$, has two orbit types: the origin (Type IaIIbIII0), and its complement (Type IaIIbIIIb). A cross section for the latter type is $W_3 = \{s \in \mathbb{C}\ ;\ s > 0\}$.

<u>Type IaIIbIIIa.</u>

We say a Type IaIIb curve $f\colon S \to M$ is of Type IaIIbIIIa if for some, hence any, second order frame field u along f we have $u^*(\omega_2^2 + 3\omega_3^3) = 0$. Since the isotropy subgroup of G_2 at the origin of $\mathbb{C}$ is all of G_2, it follows that $L_3 = L_2$. Since $z_{;1} = sz = 0$, it follows that z is constant, and so there are no non-constant invariants.

<u>Theorem</u>. Any Type IaIIbIIIa curve is G-congruent to an open submanifold of the homogeneous, Type IaIIbIIIa curve $h\colon \mathbb{C}P^1 \to \mathbb{C}G_{4,2}$ defined

$$\text{by}\quad h\begin{bmatrix} a \\ b \end{bmatrix} = \begin{bmatrix} a^2 & 0 \\ 0 & 1 \\ \sqrt{2}ab & 0 \\ b^2 & 0 \end{bmatrix}, \quad a, b \in \mathbb{C} .$$

<u>Proof</u>: Let $u\colon S \to G$ be a second order frame along f. By hypothesis, $u^*\omega_2^3 = u^*\omega_1^4 = u^*\omega_2^4 = u^*\omega_2^1 = u^*(\omega_2^2 + 3\omega_3^3) = 0$ and $u^*\omega_3^4 = zu^*\omega_1^3$, where z is a positive constant. We first observe that z must be equal to 1. In fact, taking the exterior derivative of $u^*(\omega_2^2 + 3\omega_3^3) = 0$, and applying the Maurer-Cartan equations of G, it follows that $0 = du^*(\omega_2^2 + 3\omega_3^3) = (1 - z^2)u^*(\omega_1^3 \wedge \bar{\omega}_1^3)$. Hence $z = 1$, since $u^*\omega_1^3$ is a nowhere vanishing $(1, 0)$ form.

Let $\mathfrak{h}$ denote the 4-dimensional, left-invariant, involutive distribution on G defined by the equations $\{\omega_2^3 = 0$, $\omega_1^4 = 0$, $\omega_2^4 = 0$, $\omega_2^1 = 0$, $\omega_3^4 - \omega_1^3 = 0$, $\omega_2^2 + 3\omega_3^3 = 0$. Let H be the maximal connected integral submanifold of $\mathfrak{h}$ through the identity, i.e., the analytic subgroup of G with Lie algebra $\mathfrak{h}$. Since $u(S)$ is an integral submanifold of $\mathfrak{h}$, a G-translate of $u(S)$ must like in H. Translating $f(S)$ by the same, we may as well assume $u(S) \subset H$.

$$\text{Now}\quad \mathfrak{h} = \left\{ \begin{pmatrix} it & 0 & -\bar{w} & 0 \\ 0 & -3ir & 0 & 0 \\ w & 0 & ir & -\bar{w} \\ 0 & 0 & w & i(2r-t) \end{pmatrix} : w \in \mathbb{C},\ t, r \in \mathbb{R} \right\} = \mathfrak{h}_1 \oplus \mathcal{C},$$

where $\mathcal{C}$ is the center $\left\{\begin{pmatrix} it & & & \\ & -3it & & \\ & & it & \\ & & & it \end{pmatrix} : t \in \mathbb{R}\right\}$, and

$\mathfrak{g}_1 \cong su(2)$. In fact, let $\rho\colon U(2) \to SU(4)$ be the representation given by: for $\begin{pmatrix} c & 0 \\ 0 & c \end{pmatrix}$ in the center of $U(2)$, $\rho\begin{pmatrix} c & 0 \\ 0 & c \end{pmatrix}$

$$= \begin{pmatrix} c & & & \\ & \bar{c}^3 & & \\ & & c & \\ & & & c \end{pmatrix}, \text{ and for } A = \begin{pmatrix} a & -\bar{b} \\ b & \bar{a} \end{pmatrix} \in SU(2),$$

$$\rho(A) = \begin{pmatrix} a^2 & 0 & -\sqrt{2}a\bar{b} & \bar{b}^2 \\ 0 & 1 & 0 & 0 \\ \sqrt{2}ab & 0 & a\bar{a}-b\bar{b} & -\sqrt{2}\bar{a}\bar{b} \\ b^2 & 0 & \sqrt{2}\bar{a}b & \bar{a}^2 \end{pmatrix}.$$ Then $\rho_* su(2) = \mathfrak{g}_1$ and

ρ_* (center of $u(2)$) $= \mathcal{C}$. Furthermore, for $A \in U(2)$ and $\begin{bmatrix} v \\ w \end{bmatrix} \in \mathbb{C}P^1$, $h(A\begin{bmatrix} v \\ w \end{bmatrix}) = \rho(A)h\begin{bmatrix} v \\ w \end{bmatrix}$. Hence $f(S) \subset \pi(H) = \pi(\rho U(2))$

$$= \rho U(2)\cdot\begin{bmatrix} 1 & 0 \\ 0 & 1 \\ 0 & 0 \\ 0 & 0 \end{bmatrix} = \rho U(2)\cdot h\begin{bmatrix} 1 \\ 0 \end{bmatrix} = h(U(2)\cdot\begin{bmatrix} 1 \\ 0 \end{bmatrix}) = h\mathbb{C}P^1 .$$

To see that h is, in fact, a Type IaIIbIIIa curve, construct a Frenet frame along h using ρ in the usual way.

Type IaIIbIIIb.

We say a Type IaIIb curve $f\colon S \to M$ is of Type IaIIbIIIb if there is a second order frame field u along f such that $s > 0$, where $u^*(\omega_2^2 + 3\omega_3^3) = s\theta - \bar{s}\bar{\theta}$. This condition characterizes the third order frame fields along f, and $s\colon S \to \mathbb{R}^+$ is a third order invariant.

The isotropy subgroup of G_2 at each point of W_3 is G_3

$$= \left\{ \begin{pmatrix} a & & & \\ & \bar{a}^3 & & \\ & & a & \\ & & & a \end{pmatrix} : a \in U(1) \right\}.$$ Its Lie algebra is $\mathfrak{g}_3$

$$= \left\{ \begin{pmatrix} i & & & \\ & -3i & & \\ & & i & \\ & & & i \end{pmatrix} t : t \in \mathbb{R} \right\},$$ and for a subspace of $\mathfrak{g}_2$ complementary to $\mathfrak{g}_3$ we take $\mathfrak{m}_3 = \left\{ \begin{pmatrix} i & & & \\ & 0 & & \\ & & 0 & \\ & & & -i \end{pmatrix} t : t \in \mathbb{R} \right\}$. The decomposition $\mathfrak{g}_2 = \mathfrak{g}_3 + \mathfrak{m}_3$ decomposes ω_2 into $\omega_2 = \omega_3 + \Theta_3$, where ω_3 is the $\mathfrak{g}_3$-component and $\Theta_3 = \frac{1}{2}\left(\omega_1^1 - \omega_4^4\right)\begin{pmatrix} 1 & & & \\ & 0 & & \\ & & 0 & \\ & & & -1 \end{pmatrix}$.

Let u be a third order frame field along f, let $u^*\omega_1^3 = \theta$ so $u^*\omega_3^4 = z\theta$ and $u^*(\omega_2^2 + 3\omega_3^3) = s(\theta - \bar{\theta})$ where $z, s\colon S \to \mathbb{R}^+$. Let $u^*(\omega_1^1 - \omega_4^4)$ $\xi\theta - \bar{\xi}\theta$, where $\xi\colon S \to \mathbb{C}$ is a smooth function.

Any other third order frame field along f is given by $\mathfrak{U} = uK$, where

$$K = \begin{pmatrix} a & & & \\ & \bar{a}^3 & & \\ & & a & \\ & & & a \end{pmatrix}$$, $a: S \to U(1)$ a smooth map. Let $\tilde{\theta}, \tilde{\xi}$ be the analogs of θ, ξ for $\tilde{u}$. Then $\tilde{u}^*(\Theta_0 + \ldots + \Theta_3) = Ad(K^{-1})u^*(\Theta_0 + \ldots + \Theta_3)$ $= u^*(\Theta_0 + \ldots + \Theta_3)$ implies that $\tilde{\theta} = \theta$, $\tilde{\xi} = \xi$ and hence all higher order frames are the same as the third order frames, i.e., $L_j = L_3$ for all $j \geqslant 3$. The function ξ is a fourth order invariant.

The structure equations of f are:

$$d\theta = \frac{1}{2}(\bar{\xi} - s)\theta \wedge \bar{\theta}\ , \quad dz = sz(\theta + \bar{\theta})\ ,$$

$$2\text{Re}\ s_{;1} = 3(z^2 - 1) - s^2 + s\text{Re}\ \xi\ ,$$

and

$$2\text{Re}\ \xi_{;2} = 1 + z^2 + |\xi|^2 - s\text{Re}\ \xi\ ,$$

where $z > 0$ is the second order invariant, $s > 0$ is a third order invariant, ξ is a $\mathbb{C}$-valued fourth order invariant, $ds = s_{;1}\theta + \bar{s}_{;1}\bar{\theta}$, and $d\xi = \xi_{;1}\theta + \xi_{;2}\bar{\theta}$.

Furthermore, from the equations $0 = ddz$ and $0 = dds$, it follows that $2\ \text{Im}\ s_{;1} = s\ \text{Im}\ \xi$ and $s\ \text{Im}\ \xi_{;2} = -(3(z^2 - 1) + \frac{s^2}{2})\ \text{Im}\ \xi$. Thus "Re" can be removed from both sides of the third structure equation.

Now $z_{;1} = sz > 0$ implies z is non-constant, and it is easily verified that $dz \wedge ds = 0$ iff $\text{Im}\ \xi = 0 = \text{Im}\ s_{;1}$. Thus, either z and s are independent, or all invariants are real valued functions of z .

The fine structure has two parts:

Type IaIIbIIIbi, which is the case when z and s are independent; and

Type IaIIbIIIbii, which is the case when all invariants are real valued functions of z. In the latter case there is a 2-parameter subgroup of G acting on $f(S)$ by translations along the z = constant curves of S.

Type IaIIc.

We say a Type Ia curve $f: S \to M$ is of Type IaIIc if for some, hence any, first order frame field u along f the forms $u^*\omega_2^1$ and $u^*\omega_3^4$ never vanish. A cross section for the action of G_1 is then $W_2 = \{(y, z) \in \mathbb{C}^2 : y > 0 , z > 0\}$. There exists (at least locally) a first order frame field u along f such that $y > 0$ and $z > 0$, where, we recall, $u^*\omega_2^1 = y\theta$, $u^*\omega_3^4 = z\theta$ and $\theta = u^*\omega_1^3$. This condition characterizes the second order frame fields along f. The smooth positive functions y and z are the second order invariants.

The isotropy subgroup of G_1 at any point of W_2 is G_2

$$= \left\{ \begin{pmatrix} a & & & \\ & \varepsilon a^3 & & \\ & & \varepsilon\bar{a} & \\ & & & \bar{a}^3 \end{pmatrix} : a \in U(1) , \varepsilon^2 = 1 \right\},$$

which has two connected components, corresponding to two orientation classes of second order frames on f. Its Lie algebra is

$$\mathfrak{g}_2 = \left\{ \begin{pmatrix} i & & & \\ & 3i & & \\ & & -i & \\ & & & -3i \end{pmatrix} t : t \in \mathbb{R} \right\},$$

and for a subspace of $\mathfrak{g}_1$ complementary to $\mathfrak{g}_2$ we take the

$$\mathrm{Ad}(G_2)\text{-invariant subspace } \mathfrak{m}_2 = \left\{ \begin{pmatrix} ir & & & \\ & is & & \\ & & -i(2r+3s) & \\ & & & i(r+2s) \end{pmatrix} : r, s \in \mathbb{R} \right\} .$$

Then ω_1, the $\mathfrak{g}_1$-component of Ω, decomposes into $\omega_1 = \omega_2 + \Theta_2$,

$$\text{where the } \mathfrak{m}_2\text{-component is } \Theta_2 = \frac{1}{10} \begin{pmatrix} 4\sigma+\tau & 0 & 0 & 0 \\ 0 & -3\sigma-2\tau & 0 & 0 \\ 0 & 0 & \sigma+4\tau & 0 \\ 0 & 0 & 0 & -2\sigma-3\tau \end{pmatrix},$$

$\sigma = 3\omega_1^1 + \omega_4^4$, $\tau = \omega_2^2 + 3\omega_3^3$; and ω_2 is the $\mathfrak{g}_2$-component.

Let u be a second order frame field along f, let $u^*\omega_1^3 = \theta$, $u^*\omega_2^1 = y\theta$ and $u^*\omega_3^4 = z\theta$, where $y, z: S \to \mathbb{R}^+$; and set $u^*(3\omega_1^1 + \omega_4^4) = r\theta - \bar{r}\bar{\theta}$ and $u^*(3\omega_3^3 + \omega_2^2) = s\theta - \bar{s}\bar{\theta}$, where $r, s: S \to \mathbb{C}$ are smooth functions. Taking the exterior derivatives of $u^*\omega_2^1 = y\theta$ and $u^*\omega_3^4 = z\theta$, and applying the Maurer-Cartan equations of G, it follows that $y_{;1} = yr$ and $z_{;1} = zs$.

Let $\tilde{u} = uK$ be any other second order frame field along f,

$$\text{where } K = \begin{pmatrix} a & 0 & 0 & 0 \\ 0 & \varepsilon a^3 & 0 & 0 \\ 0 & 0 & \varepsilon\bar{a} & 0 \\ 0 & 0 & 0 & \bar{a}^3 \end{pmatrix}, \quad a: S \to U(1), \quad \varepsilon^2 = 1 .$$

Let $\tilde{\theta}$, $\tilde{r}$, $\tilde{s}$ be the analogues of θ, r, s for $\tilde{u}$. Then $\tilde{u}^*(\Theta_0 + \Theta_1 + \Theta_2) = \mathrm{ad}(K^{-1})u^*(\Theta_0 + \Theta_1 + \Theta_2)$ implies that $\tilde{\theta} = \varepsilon a^2\theta$, $\tilde{u}^*(3\omega_1^1 + \omega_4^4) = \tilde{u}^*(3\omega_1^1 + \omega_4^4)$, $u^*(\omega_2^2 + 3\omega_3^3) = u^*(\omega_2^2 + 3\omega_3^3)$, $\tilde{r} = \varepsilon\bar{a}^2 r$, and $\tilde{s} = \varepsilon\bar{a}^2 s$.

Thus we have the action of G_2 on $\mathbb{C}^2$ given by $(a, \varepsilon)(r, s) = (\varepsilon a^2 r, \varepsilon a^2 s)$, where $a \in U(1)$, $\varepsilon = \pm 1$, $r, s \in \mathbb{C}$. There are two orbit types: the origin (Type IaIIcIIIa) and its complement (Type IaIIcIIIb). A local cross section for the latter orbits is $W_2 = \{(r, s) \in \mathbb{C}^2 : r > 0\}$.

Type IaIIcIIIa.

We say a Type IaIIc curve $f: S \to M$ is of Type IaIIcIIIa if for some, hence any second order frame field u along f we have $u^*(3\omega_1^1 + \omega_4^4) = 0$ and $u^*(3\omega_3^3 + \omega_2^2) = 0$, i.e., $r = s = 0$. Since the action of G_2 on the origin of $\mathbb{C}^2$ is trivial, all higher order frames are the same as the second order frames; i.e., $L_j = L_2$ for all $j \geqslant 2$. There are no third order invariants. Furthermore, $y_{;1} = yr = 0$ and $z_{;1} = zs = 0$ implies y and z are constant, so there are no non-constant invariants. Taking the exterior derivative of $u^*(3\omega_1^1 + \omega_4^4) = 0$ and $u^*(\omega_2^2 + 3\omega_3^3) = 0$, and applying the Maurer-Cartan equations of G, it follows that $y = z = \frac{\sqrt{3}}{2}$. The Frenet frames have order 2.

Theorem. Any curve of Type IaIIcIIIa is G-congruent to an open submanifold of the homogeneous, Type IaIIcIIIa curve $h: \mathbb{C}P^1 \to M$ defined by

$$h\begin{bmatrix} v \\ w \end{bmatrix} = \begin{bmatrix} v^2\bar{v} - 2vw\bar{w} & \sqrt{3}\, v^2 w \\ -\sqrt{3}\, v^2\bar{w} & v^3 \\ 2v\bar{v}w - w^2\bar{w} & \sqrt{3}\, vw^2 \\ \sqrt{3}\, \bar{v}\, w^2 & w^3 \end{bmatrix}, \quad v, w \in \mathbb{C}.$$

Proof: Let $\mathfrak{h}$ be the 3-dimensional, left-invariant, involutive distribution on G defined by the equations $\{\omega_2^3 = 0\ ,\ \omega_1^4 = 0\ ,\ \omega_2^4 = 0\ ,\ \omega_2^1 - \frac{\sqrt{3}}{2}\omega_1^3 = 0\ ,\ \omega_3^4 - \frac{\sqrt{3}}{2}\omega_1^3 = 0\ ,\ 3\omega_1^1 + \omega_4^4 = 0\ ,\ \omega_2^2 + 3\omega_3^3 = 0\}$. Let H be the analytic subgroup of G with Lie algebra $\mathfrak{h}$. Since $u(S)$ is an integral manifold of $\mathfrak{h}$, (for any second order frame u along f), a G-translate of $u(S)$ is contained in H. Translating f, we may as well assume $u(S) \subset H$.

Now H is the degree 4 representation ρ of $SU(2)$ on the symmetric algebra $\mathcal{S}_3(\mathbb{C}^2)$ with respect to the ordered basis $\sqrt{3}\varepsilon_1^2\varepsilon_2$, ε_1^3, $\sqrt{3}\,\varepsilon_1\varepsilon_2^2$, ε_2^3, where $\varepsilon_1, \varepsilon_2$ is the standard basis of $\mathbb{C}^2$. In fact, one needs only check that $\mathfrak{h} = \rho_* su(2)$, where for $X = \begin{pmatrix} it & -\bar{w} \\ w & -it \end{pmatrix} \in su(2)$, $(\rho_* X)(\varepsilon_i\varepsilon_j\varepsilon_k) = (X\varepsilon_i)\varepsilon_j\varepsilon_k + \varepsilon_i(X\varepsilon_j)\varepsilon_k + \varepsilon_i\varepsilon_j(X\varepsilon_k)$.

Furthermore, if $A \in SU(2)$, $\begin{bmatrix} v \\ w \end{bmatrix} \in \mathbb{C}P^1$, then $h(A\begin{bmatrix} v \\ w \end{bmatrix}) = \rho(A)h\begin{bmatrix} v \\ w \end{bmatrix}$. Hence $f(S) = \pi u(S) \subset \pi H = \pi\rho(SU(2)) = \rho(SU(2))\cdot h\begin{bmatrix} 1 \\ 0 \end{bmatrix} = h(SU(2)\cdot\begin{bmatrix} 1 \\ 0 \end{bmatrix}) = h\mathbb{C}P^1$.

In order to show that h is in fact a Type IaIIcIIIa holomorphic curve, we construct a zeroth order frame field u along h, defined say on the open neighborhood $\{\begin{bmatrix} z \\ w \end{bmatrix} : z \neq 0\} \subseteq \mathbb{C}P^1$, by $u\begin{bmatrix} 1 \\ z \end{bmatrix} = \rho c\begin{pmatrix} 1 & -\bar{z} \\ z & 1 \end{pmatrix}$, where $c = (1 + |z|^2)^{-\frac{1}{2}}$. Then $u^*\Omega$ is an $\mathfrak{h}$-valued 1-form on $\mathbb{C}P^1$, since the image of u is contained in H. Hence u is actually a Frenet frame and h is a Type IaIIcIIIa holomorphic curve.

Type IaIIcIIIb.

We say a Type IaIIc curve $f: S \to M$ is of Type IaIIcIIIb if there exists a second order frame field u along f such that $r > 0$, where $u^*(3\omega_1^1 + \omega_4^4) = r\theta - \overline{r\theta}$. This condition characterizes the third order frame fields. (We shall not discuss the similar case when for all second order frame fields u along f, $u^*(3\omega_1^1 + \omega_4^4) = 0$ but $u^*(3\omega_3^3 + \omega_2^2) \neq 0$, in which case $s > 0$ is an appropriate condition for defining the third order frame fields.)

The smooth functions $r > 0$ and $s: S \to \mathbb{C}$ are third order invariants. The isotropy subgroup of G_2 at any point of W_2 is $G_3 = \{\pm I\}$ All higher order frames are the same as the third order frames; $L_j = L_3$ for all $j \geqslant 3$. Furthermore $\Theta_3 = \omega_2 = \frac{\nu}{10}\begin{pmatrix} 1 & & & \\ & 3 & & \\ & & -3 & \\ & & & -1 \end{pmatrix}$ where $\nu = \omega_1^1 + 2\omega_2^2 - \omega_3^3$. For any third order frame field u along f, we set $u^*\nu = \beta\theta - \overline{\beta\theta}$, so the smooth function $\beta: S \to \mathbb{C}$ is a fourth order invariant. The form θ is a third order invariant form.

In summary, the third order frames on f are uniquely determined up to two orientation classes. If u is a third order frame field along f then $u^*\omega_1^3 = \theta$; $u^*\omega_2^1 = y\theta$, $u^*\omega_3^4 = z\theta$ where $y > 0$, $z > 0$ are second order invariants; $u^*(3\omega_1^1 + \omega_4^4) = r(\theta - \overline{\theta})$, $u^*(3\omega_3^3 + \omega_2^2) = s\theta - \overline{s\theta}$ where $r > 0$, $s: S \to \mathbb{C}$ are third order invariants; and $u^*(\omega_1^1 + 2\omega_2^2 - \omega_3^3) = \beta\theta - \overline{\beta\theta}$ where $\beta: S \to \mathbb{C}$ is a fourth order invariant. The structure equations of f are:

$$d\theta = \frac{1}{8}(3r - \bar{s} + 2\bar{\beta})\theta \wedge \bar{\theta}\ , \quad y_{;1} = yr\ , \quad z_{;1} = zs\ ,$$

$$2\mathrm{Re}\ r_{;1} = 3(1 - y^2) - z^2 + \frac{3}{4} r^2 - \frac{1}{4} r\mathrm{Re}\ s + \frac{1}{2} r\mathrm{Re}\ \beta\ ,$$

$$2\mathrm{Re}\ s_{;2} = y^2 + 3z^2 - 3 - \frac{1}{4} |s|^2 + \frac{3}{4} r\mathrm{Re}\ s + \frac{1}{2} \mathrm{Re}\ s\bar{\beta}\ ,$$

and

$$2\mathrm{Re}\ \beta_{;2} = y^2 - z^2 + 2 + \frac{1}{2} |\beta|^2 + \frac{3}{4} r\mathrm{Re}\ \beta - \frac{1}{4} \mathrm{Re}\ \beta\bar{s}\ ,$$

Computing $0 = ddy$ and $0 = ddz$, it follows that the fourth and fifth structure equations remain valid with the "Re" omitted everywhere. Notice that y is non-constant.

There are three possibilities in the fine structure:

Type IaIIcIIIbi: y and z are independent, Frenet frames have order 3;

Type IaIIcIIIbii: y and z are dependent, (i.e., $\mathrm{Im}\ s = 0$) , but y and r are independent, (i.e., $\mathrm{Im}\ \beta \neq 0$) ; Frenet frames have order 3;

Type IaIIcIIIbiii: All invariants are a function of y , and consequently all invariants are real; Frenet frames have order 3. A 1-parameter subgroup H of G acts on $f(S)$ by translating along the y = constant curves on S .

This completes the analysis of Type Ia curves.

Type Ib.

Fortunately, for the sake of conservation of paper, there is essentially only one holomorphic curve of Type Ib in M .

We say $f: S \to M$ is of Type Ib if there is a zeroth order frame field u along f for which $u^*\omega_2^3 = 0 = u^*\omega_1^4$ and $u^*\omega_2^4 = u^*\omega_1^3$. This condition characterizes the first order frame fields along f. There are no first order invariants.

The isotropy subgroup of G_0 at $\begin{bmatrix} 1 & 0 \\ 0 & 1 \end{bmatrix}$ is G_1 $= \left\{ \begin{pmatrix} A & \\ & aA \end{pmatrix} : A \in U(2) \ , \ a = \pm \det A^{-1} \right\}$, which has two connected components, corresponding to two orientation classes of first order frames.

Its Lie algebra is
$$\mathfrak{g}_1 = \left\{ \begin{pmatrix} ir & -\bar{z} & & \\ z & is & & 0 \\ & & -is & -\bar{z} \\ 0 & & z & -ir \end{pmatrix} : r, s \in \mathbb{R} \ , \ z \in \mathbb{C} \right\} .$$

For a subspace of $\mathfrak{g}_0$ complementary to $\mathfrak{g}_1$ we take the $Ad(G_1)$-invariant subspace
$$\mathfrak{m}_1 = \left\{ \begin{pmatrix} it & -\bar{z} & & \\ z & -it & & 0 \\ & & -it & \bar{z} \\ 0 & & -z & it \end{pmatrix} : t \in \mathbb{R} \ , \ z \in \mathbb{C} \right\} .$$
The decomposition $\mathfrak{g}_0 = \mathfrak{g}_1 + \mathfrak{m}_1$ decomposes ω_0 into $\omega_0 = \omega_1 + \theta_1$,

where ω_1 is the $\mathfrak{g}_1$-component and
$$\theta_1 = \begin{pmatrix} \alpha & -\bar{\beta} & & \\ \beta & -\alpha & & 0 \\ & & -\alpha & \bar{\beta} \\ 0 & & -\beta & \alpha \end{pmatrix} ,$$
$\alpha = \frac{1}{2} (\omega_1^1 + \omega_4^4)$ and $\beta = \frac{1}{2} (\omega_1^2 - \omega_3^4)$.

Let u be a first order frame field along f and set $u^*\omega_1^3 = \theta$ so $u^*\omega_2^4 = \theta$, $u^*\omega_2^3 = 0 = u^*\omega_1^4$. Taking the exterior derivative of these last three equations, and applying the Maurer-Cartan equations of G, it follows that $u^*\alpha = 0 = u^*\beta$. It follows that all higher order frames are the same as the first order frames; i.e., $L_j = L_1$ for all $j \geq 1$. There are no non-constant invariants. The Frenet frames have first order.

<u>Theorem</u>. Any holomorphic curve $f: S \to M$ of Type Ib is G-congruent to an open submanifold of the homogeneous, Type Ib curve $h: \mathbb{C}P^1 \to M$ defined by $h\begin{bmatrix} v \\ w \end{bmatrix} = \begin{bmatrix} vI \\ wI \end{bmatrix}$, where I is the 2×2 identity matrix.

<u>Proof</u>: Consider the 6-dimensional, left-invariant, involutive distribution $\mathfrak{h}$ defined on G by the equations $\{\omega_2^3 = 0 , \ \omega_1^4 = 0 , \ \omega_2^4 = \omega_1^3 , \ \omega_2^1 = \omega_4^3 , \ \omega_1^1 + \omega_4^4 = 0\}$. It is easily checked that the analytic subgroup of G with Lie algebra $\mathfrak{h}$ is

$$H = \left\{\begin{pmatrix} Ba & -B\bar{b} \\ Bb & B\bar{a} \end{pmatrix} : B \in SU(2) , \ a\bar{a} + b\bar{b} = 1\right\},$$ which is isomorphic with $SU(2) \times SU(2)$.

Let u be a first order frame field along f. Then $u(S)$ is an integral submanifold of $\mathfrak{h}$, and thus a G-translate of $u(S)$ is contained in H. Translating f, we may as well assume $u(S) \subset H$.

Then $$f(S) = \pi u(S) \subset \pi H = H \cdot \begin{bmatrix} I \\ 0 \end{bmatrix} = \left\{\begin{bmatrix} Ba \\ Bb \end{bmatrix} : B \in SU(2) , \ a\bar{a} + b\bar{b} = 1\right\}$$

$$= \left\{\begin{bmatrix} aI \\ bI \end{bmatrix} : a\bar{a} + b\bar{b} = 1\right\} = h\mathbb{C}P^1 .$$

To see that h is, in fact, a Type Ib curve, we construct a zeroth order frame u along h, defined, say, on the open neighborhood

$\left\{\begin{bmatrix} z \\ w \end{bmatrix} : w \neq 0\right\} \subseteq \mathbb{C}P^1$, by: $u\begin{bmatrix} 1 \\ z \end{bmatrix} = c\begin{pmatrix} I & -\bar{z}I \\ zI & I \end{pmatrix} \in SU(4)$, where

$c = (1 + |z|^2)^{-\frac{1}{2}}$ and $I = \begin{pmatrix} 1 & 0 \\ 0 & 1 \end{pmatrix}$. Then the image of u is contained in H , which implies that $u^*\Omega$ is an $\mathfrak{h}$-valued 1-form on $\mathbb{C}P^1$. Hence u is actually a Frenet frame along h and h is Type Ib.

Type Ic.

We say $f: S \to M$ is of Type Ic if there is a zeroth order frame field u along f for which $u^*\omega_1^4 = u^*\omega_2^3 = 0$ and $u^*\omega_2^4 = ku^*\omega_1^3$ where k is a smooth function on S satisfying $0 < k < 1$. This condition characterizes the first order frame fields along f , and k is the first order invariant.

The isotropy subgroup of G_0 at any point of W_1 is G_1

$$= \left\{\begin{pmatrix} a & 0 & & \\ 0 & b & & 0 \\ & & \varepsilon b^{-1} & 0 \\ & 0 & 0 & \varepsilon a^{-1} \end{pmatrix} : a, b \in U(1) , \varepsilon = \pm 1\right\} ,$$ which has two connected components corresponding to two orientation classes of first order frames on f . Its Lie algebra is $\mathfrak{g}_1$

$$= \left\{\begin{pmatrix} ir & & & \\ & it & & \\ & & -it & \\ & & & -ir \end{pmatrix} : r, t \in \mathbb{R}\right\} ,$$ and for a subspace of $\mathfrak{g}_0$ complementary to $\mathfrak{g}_1$ we take the $Ad(G_1)$-invariant subspace

$$\mathfrak{m}_1 = \left\{ \begin{pmatrix} it & -\bar{z} & & \\ z & -it & & 0 \\ & & -it & -\bar{w} \\ 0 & & w & it \end{pmatrix} : t \in \mathbb{R} \ , \ z \in \mathbb{C} \right\} .$$ The decomposition

$\mathfrak{g}_0 = \mathfrak{g}_1 + \mathfrak{m}_1$ decomposes ω_0 into $\omega_0 = \omega_1 + \Theta_1$, where ω_1 is the $\mathfrak{g}_1$-component and $\Theta_1 = \begin{pmatrix} \nu & \omega_2^1 & & 0 \\ \omega_1^2 & -\nu & & \\ & & -\nu & \omega_4^3 \\ 0 & & \omega_3^4 & \nu \end{pmatrix}$, where $\nu = \frac{1}{2}(\omega_1^1 + \omega_4^4)$.

If u is a first order frame field along f, we set $u^*\omega_1^3 = \theta$, so $u^*\omega_2^4 = k\theta$ and $u^*\omega_2^3 = 0 = u^*\omega_1^4$. Taking the exterior derivative of these last three equations, and applying the Maurer-Cartan equations of G, it follows that: $u^*(\omega_2^1 - k\omega_4^3)$ and $u^*(\omega_3^4 - k\omega_1^2)$ are bidegree $(1, 0)$ forms on S, and $\theta \wedge (\bar{k}_{;1}\bar{\theta} + 2k(\omega_1^1 + \omega_4^4)) = 0$, where $dk = k_{;1}\theta + \bar{k}_{;1}\bar{\theta}$. Thus we may set $u^*(\omega_2^1 - k\omega_4^3) = y\theta$ and $u^*(\omega_3^4 - k\omega_1^2) = z\theta$ for some smooth function $y, z\colon S \to \mathbb{C}$; and if we set $u^*(\omega_1^1 + \omega_4^4) = x\theta - \overline{x\theta}$ where $x\colon S \to \mathbb{C}$, then $k_{;1} = 2kx$.

Any other first order frame field along f is given by $\tilde{u} = uK$,

where $K = \begin{pmatrix} a & & & \\ & b & & \\ & & \varepsilon b^{-1} & \\ & & & \varepsilon a^{-1} \end{pmatrix}$, $\varepsilon = \pm 1$, $a, b\colon S \to U(1)$ smooth maps. Let $\tilde{\theta}, \tilde{x}, \tilde{y}, \tilde{z}$ denote the analogs of θ, x, y, z for $\tilde{u}$. Then $\tilde{u}^*(\Theta_0 + \Theta_1) = \mathrm{ad}(K^{-1})u^*(\Theta_0 + \Theta_1)$ implies that

$\tilde{u}^*(\omega_1^1 + \omega_4^4) = u^*(\omega_1^1 + \omega_4^4)$, $\tilde{\Theta} = \varepsilon ab\theta$, $\tilde{y} = \varepsilon a^{-2}y$, $\tilde{z} = \varepsilon b^{-2}z$, and $\tilde{x} = \varepsilon(ab)^{-1}x$.

The action of G_1 on $\mathbb{C}^3$ given by $(a, b, \varepsilon)(x, y, z) = \varepsilon(abx, a^2y, b^2z)$, where $\varepsilon = \pm 1$, $a, b \in U(1)$, $x, y, z \in \mathbb{C}$, has five orbit types: the origin $\{x = y = z = 0\}$ (cannot occur); the x-axis $\{y = z = 0 , x \neq 0\}$ (Type IcIIa); the y-axis $\{x = z = 0 , y \neq 0\}$ or the z-axis $\{x = y = 0 , z \neq 0\}$ (Type IcIIb); the yz-plane $\{x = 0 , y \neq 0$ and $z \neq 0\}$ (Type IcIIc); and the set $\{x \neq 0 , y \neq 0$ or $z \neq 0\}$ (Type IcIId).

The first case does not occur, because if $x = y = z = 0$ for some, hence any, first order frame field u along f , then taking the exterior derivative of $u^*(\omega_1^1 + \omega_4^4) = 0$, and applying the Maurer-Cartan equations of G together with the fact that $u^*\omega_2^1 = 0 = u^*\omega_3^4$, it follows that $k^2 - 1 = 0$, which cannot hold for a Type Ic curve. Hence x, y and z cannot all vanish.

Type IcIIa.

We say a Type Ic curve $f: S \to M$ is of Type IcIIa if for some, hence any, first order frame field u along f we have $y = z = 0$ and $x \neq 0$. Then there exists, at least locally, a first order frame field for which $x > 0$. This condition characterizes the second order frame fields along f , and the positive function x is a second order invariant of f .

The isotropy subgroup of G_1 at any point $(x, 0, 0)$ for $x > 0$

is $G_2 = \left\{ \begin{pmatrix} a & & & \\ & \varepsilon a^{-1} & & \\ & & a & \\ & & & \varepsilon a^{-1} \end{pmatrix} : a \in U(1) \right\}$, $\varepsilon = \pm 1$, which has two

connected components, corresponding to two orientation classes of second order frames on f . Its Lie algebra is $\mathfrak{g}_2$

$= \left\{ \begin{pmatrix} i & & & \\ & -i & & \\ & & i & \\ & & & -i \end{pmatrix} t : t \in \mathbb{R} \right\}$, and for a subspace of $\mathfrak{g}_1$ complementary to $\mathfrak{g}_2$ we take the $Ad(G_2)$-invariant subspace $\mathfrak{m}_2$

$= \left\{ \begin{pmatrix} i & & & \\ & i & & \\ & & -i & \\ & & & -i \end{pmatrix} t : t \in \mathbb{R} \right\}$. The decomposition $\mathfrak{g}_1 = \mathfrak{g}_2 + \mathfrak{m}_2$

decomposes ω_1 into $\omega_1 = \omega_2 + \Theta_2$, where ω_2 is the $\mathfrak{g}_2$-component

and $\Theta_2 = \alpha \begin{pmatrix} 1 & & & \\ & 1 & & \\ & & -1 & \\ & & & -1 \end{pmatrix}$, $\alpha = \frac{1}{2} (\omega_1^1 + \omega_2^2)$.

Let u be any second order frame field along f , set $u^*\omega_1^3 = \theta$, so $u^*\omega_2^4 = k\theta$, $0 < k < 1$, $u^*(\omega_2^1 - k\omega_4^3) = 0 = u^*(\omega_3^4 - k\omega_1^2)$, and $u^*(\omega_1^1 + \omega_4^4) = x(\theta - \bar{\theta})$, $x > 0$. From the third and fourth of these equations it follows that $u^*\omega_1^2 = 0 = u^*\omega_3^4$. Taking the exterior derivative of these last two equations and applying the Maurer-Cartan equations of G yields no further relations. If we set $u^*(\omega_1^1 + \omega_2^2) = \zeta\theta - \overline{\zeta\theta}$, where $\zeta\colon S \to \mathbb{C}$ is a smooth function, and take the exterior derivative of the fifth equation above, it follows that $2 \operatorname{Re} x_{;1} = 1 - k^2 + 2x^2 + 2x \operatorname{Re} \zeta$.

Any other second order frame field along f is given by $\tilde{u} = uK$,

where $K = \begin{pmatrix} a & & & \\ & \varepsilon a^{-1} & & \\ & & \varepsilon a & \\ & & & a^{-1} \end{pmatrix}$, $\varepsilon = \pm 1$, $a\colon S \to U(1)$ a smooth function. Let $\tilde{\theta}, \tilde{\zeta}$ be the analogs of θ, ζ for $\tilde{u}$. Then $\tilde{u}^*(\Theta_0 + \Theta_1 + \Theta_2) = \operatorname{ad}(K^{-1})u^*(\Theta_0 + \Theta_1 + \Theta_2) = u^*(\Theta_0 + \Theta_1 + \Theta_2)$, which implies $\tilde{\theta} = \theta$ and $\tilde{\zeta} = \zeta$. Hence all higher order frames are the same as the second order frames; i.e., $L_j = L_2$ for all $j \geq 2$. The smooth functions $\zeta\colon S \to \mathbb{C}$ is a third order invariant. The $(1, 0)$ form θ is a second order invariant form on S; i.e., independent of the choice of second order frame field along f.

The structure equations of f are:

$$d\theta = (x + \bar{\zeta})\theta \wedge \bar{\theta}, \quad k_{;1} = 2kx,$$

$$2 \operatorname{Re} x_{;1} = 1 - k^2 + 2x^2 + 2x \operatorname{Re} \zeta,$$

and

$$2 \operatorname{Re} \zeta_{;2} = 1 + k^2 + 2|\zeta|^2 + 2x \operatorname{Re} \zeta.$$

Computing $0 = ddk$ and $0 = ddx$, it follows that $\operatorname{Im} x_{;1} = x \operatorname{Im} \zeta$ and $x \operatorname{Im} \zeta_{;2} = (x^2 + k^2 - 1) \operatorname{Im} \zeta$. Thus

$x_{;1} = \frac{1}{2}(1 - k^2) + x^2 + x\zeta$. Notice that k is nonconstant.

There are two possibilities in the fine structure:

Type IcIIai: k and x are independent, (i.e., $\operatorname{Im} \zeta \neq 0$) , Frenet frames have order 2;

Type IcIIaii: all invariants are functions of k , Frenet frames have order 2. A 2-parameter subgroup $H \supset G_2$ of G acts non-transitively on $f(S)$ by translating along k = constant curves in S .

Type IcIIb.

This category consists of a 1-parameter family of homogeneous curves. We omit discussion of the case $x = y = 0$, $z \neq 0$.

We say a Type Ic curve $f: S \to M$ is of Type IcIIb if for some, hence any, first order frame field u along f we have $x = z = 0$ and $y \neq 0$. Then there exists, at least locally, a first order frame field u along f such that $y > 0$. This condition characterizes the second order frame fields along f , and the positive function y is a second order invariant. Furthermore, since $k_{;1} = 2kx = 0$, it follows that k is constant, $0 < k < 1$.

Let u be a second order frame field along f , and set $u^*\omega_1^3 = \theta$. Then $u^*\omega_2^4 = k\theta$, k constant, $0 < k < 1$; $u^*(\omega_2^1 - k\omega_4^3) = y\theta$, $y > 0$; $u^*(\omega_3^4 - k\omega_1^2) = z\theta = 0$; and $u^*(\omega_1^1 + \omega_4^4) = x\theta - \overline{x\theta} = 0$. Taking the exterior derivative of the last of these equations, and applying the Maurer-Carton equations, it follows that $y = 1 - k^2$, and thus y is constant. Combining the middle two equations above it follows that $u^*\omega_2^1 = \theta$ and $u^*\omega_4^3 = k\theta$. Taking the exterior derivative of either of these two equations, and applying the Maurer-Cartan equations, it follows that $u^*\omega_1^1 = 0$, and thus also $u^*\omega_4^4 = 0$.

The isotropy subgroup of G_1 at $(0, y, 0)$, for any $y > 0$,

is $G_2 = \left\{ \begin{pmatrix} \varepsilon & & & \\ & b & & \\ & & \varepsilon^2 b^{-1} & \\ & & & \varepsilon \end{pmatrix} : \varepsilon, b \in U(1), \ \varepsilon^4 = 1 \right\}$, which has four connected components, corresponding to four orientation classes of second order frames on f. Its Lie algebra is $\mathfrak{g}_2 = \left\{ \begin{pmatrix} 0 & & & \\ & i & & \\ & & -i & \\ & & & 0 \end{pmatrix} t : t \in \mathbb{R} \right\}$

and as a subspace of $\mathfrak{g}_1$ complementary to $\mathfrak{g}_2$ we take the $\mathrm{Ad}(G_2)$-invariant subspace $\mathfrak{m}_2 = \left\{ \begin{pmatrix} i & & & \\ & 0 & & \\ & & 0 & \\ & & & -i \end{pmatrix} t : t \in \mathbb{R} \right\}$. The decomposition $\mathfrak{g}_1 = \mathfrak{g}_2 + \mathfrak{m}_2$ decomposes ω_1 into $\omega_1 = \omega_2 + \theta_2$,

where ω_2 is the $\mathfrak{g}_2$-component and $\theta_2 = \frac{1}{2}(\omega_1^1 - \omega_4^4) \begin{pmatrix} 1 & & & \\ & 0 & & \\ & & 0 & \\ & & & -1 \end{pmatrix}$.

We have shown above that there are no non-constant first and second order invariants, and that $u^*\theta_2 = 0$ for any second order frame field u along f . Hence all higher order frames are the same as the second order frames; i.e., $L_j = L_2$, for all $j \geq 2$. The Frenet frames have second order.

A Frenet frame u along f is characterized by: $u^*\omega_1^3 = \theta$ is a non-vanishing bidegree (1, 0) form on S , $u^*\omega_2^3 = 0$, $u^*\omega_1^4 = 0$, $u^*\omega_2^4 = k\theta$, $u^*\omega_2^1 = \theta$, $u^*\omega_4^3 = k\theta$, $u^*\omega_1^1 = 0$, and $u^*\omega_4^4 = 0$, where k is constant, $0 < k < 1$. Consider the left-invariant, 3-dimensional, involutive distribution $\mathfrak{h}_k$ defined on G by the equations $\{\omega_2^3 = 0$, $\omega_1^4 = 0$, $\omega_2^4 = k\omega_1^3$, $\omega_2^1 = \omega_1^3$, $\omega_4^3 = k\omega_1^3$, $\omega_1^1 = 0$, $\omega_4^4 = 0\}$. Then

$$\mathfrak{h}_k = \left\{ \begin{pmatrix} 0 & z & -\bar{z} & 0 \\ -\bar{z} & it & 0 & -k\bar{z} \\ z & 0 & -it & kz \\ 0 & kz & -k\bar{z} & 0 \end{pmatrix} : t \in \mathbb{R} , \ z \in \mathbb{C} \right\} ,$$

a 3-dimensional Lie subalgebra of $\mathfrak{g}$. A Frenet frame u along f is characterized by the property that $u^*\Omega$ be an $\mathfrak{h}_k$-valued 1-form on S . Let H_k be the analytic subgroup of G whose Lie algebra is $\mathfrak{h}_k$

In order to identify H_k , consider the tensor product representation of SU(2) in $GL(\mathbb{C}^2 \otimes \mathbb{C}^2)$ given by $\rho: SU(2) \to GL(\mathbb{C}^2 \otimes \mathbb{C}^2)$, $\rho(A)(\varepsilon_i \otimes \varepsilon_j) = A\varepsilon_i \otimes A\varepsilon_j$, $i, j = 1, 2$, where $\varepsilon_1, \varepsilon_2$ is the standard basis of $\mathbb{C}^2$. For each real number k between 0 and 1 , define ψ , $0 < \psi < \frac{\pi}{4}$, by $(*)$ $\tan \psi = (1 - k)(1 + k)^{-1}$. Let $\rho_k: SU(2) \to SU(4)$ be the matrix representation of ρ with respect to the orthonormal basis

$$(v_1, v_2, v_3, v_4) = (\varepsilon_1 \otimes \varepsilon_2, \varepsilon_1 \otimes \varepsilon_1, \varepsilon_2 \otimes \varepsilon_2, \varepsilon_2 \otimes \varepsilon_1)C_k ,$$

where $\mathbf{C}_k = \begin{pmatrix} \cos\psi & 0 & 0 & -\sin\psi \\ 0 & 1 & 0 & 0 \\ 0 & 0 & 1 & 0 \\ \sin\psi & 0 & 0 & \cos\psi \end{pmatrix}$. If $A = \begin{pmatrix} a & -\bar{b} \\ b & \bar{a} \end{pmatrix} \in SU(2)$,

where $a, b \in \mathbb{C}$, $|a|^2 + |b|^2 = 1$, then $\rho_k(A)$

$$= \begin{pmatrix} |a|^2 - |b|^2 \sin 2\psi & ab(\cos\psi + \sin\psi) & -\bar{a}\bar{b}(\cos\psi + \sin\psi) & -|b|^2 \cos 2\psi \\ -a\bar{b}(\cos\psi + \sin\psi) & a^2 & \bar{b}^2 & -a\bar{b}(\cos\psi - \sin\psi) \\ \bar{a}b(\cos\psi + \sin\psi) & b^2 & \bar{a}^2 & \bar{a}b(\cos\psi - \sin\psi) \\ -|b|^2 \cos 2\psi & ab(\cos\psi - \sin\psi) & -\bar{a}\bar{b}(\cos\psi - \sin\psi) & |a|^2 + |b|^2 \sin 2\psi \end{pmatrix};$$

and if $X = \begin{pmatrix} it & -\bar{z} \\ z & -it \end{pmatrix} \in su(2)$, $t \in \mathbb{R}$, $z \in \mathbb{C}$, then $\rho_{k*}X$

$$= \begin{pmatrix} 0 & z(\cos\psi + \sin\psi) & -\bar{z}(\cos\psi + \sin\psi) & 0 \\ -\bar{z}(\cos\psi + \sin\psi) & 2it & 0 & -\bar{z}(\cos\psi - \sin\psi) \\ z(\cos\psi + \sin\psi) & 0 & -2it & z(\cos\psi + \sin\psi) \\ 0 & z(\cos\psi - \sin\psi) & -\bar{z}(\cos\psi - \sin\psi) & 0 \end{pmatrix}.$$

Hence $\rho_{k*}su(2) = \mathfrak{h}_k$, and thus $\rho_k SU(2) = H_k$.

<u>Theorem</u>. Let $f\colon S \to M$ be a Type IcIIb holomorphic curve whose (constant) first order invariant is k, $0 < k < 1$. Then f is G-congruent to an open submanifold of the holomorphic curve $h_k\colon \mathbb{C}P^1 \to M$ defined by $h_k\begin{bmatrix} a \\ b \end{bmatrix}$ = the 4×2 matrix given by the first two columns of

$\rho_k\begin{pmatrix} a & -\bar{b} \\ b & \bar{a} \end{pmatrix}$, where $a, b \in \mathbb{C}$, $|a|^2 + |b|^2 = 1$, and ψ is defined by (*). Furthermore, for any number k such that $0 < k < 1$, this map $h_k\colon \mathbb{C}P^1 \to M$ is a holomorphic curve of Type IcIIb whose first order invariant is k .

Proof: Let u be a Frenet frame field along f . We have already observed that $u(S)$ is an integral submanifold of the distribution $\mathfrak{h}_k$ on G , and hence a G-translate of $u(S)$ lies in H_k . Translating f by the same amount, we may as well assume that $u(S) \subset H_k$. Then $f(S) = \pi u(S) \subset \pi H_k = H_k \cdot \begin{bmatrix} 1 \\ 0 \end{bmatrix} = \rho_k(SU(2)) \cdot h_k \begin{bmatrix} 1 \\ 0 \end{bmatrix} = h_k(SU(2) \cdot \begin{bmatrix} 1 \\ 0 \end{bmatrix})$ $= h_k \mathbb{C}P^1$.

It is an easy exercise to check directly that for any $0 < k < 1$, $h_k\colon \mathbb{C}P^1 \to \mathbb{C}G_{4,2}$ is holomorphic. A zeroth order frame u is defined along h_k , say on the neighborhood $\{[\begin{smallmatrix} z \\ w \end{smallmatrix}]\colon z \neq 0\} \subseteq \mathbb{C}P^1$, by $u\begin{bmatrix} 1 \\ z \end{bmatrix}$ $= \rho_k c\begin{pmatrix} 1 & -\bar{z} \\ z & 1 \end{pmatrix}$, where $c = (1 + |z|^2)^{-\frac{1}{2}}$. But then $u^*\Omega$ is an $\mathfrak{h}_k$-valued 1-form on $\mathbb{C}P^1$, since the image of u is contained in H_k . Hence u is a Frenet frame and h_k is a Type IcIIb curve with first order invariant equal to k .

Type IcIIc.

We say a Type Ic curve $f\colon S \to M$ is of Type IcIIc if for some, hence any, first order frame field u along f we have $x = 0$, $y \neq 0$ and $z \neq 0$. Then there exists, at least locally, a first order frame field for which $y > 0$ and $z > 0$. This condition characterizes second order frame fields along f, and the positive smooth functions y and z are second order invariants of f. Since $k_{;1} = 2kx = 0$, the first order invariant k is constant, $0 < k < 1$.

The isotropy subgroup of G_1 at any point $(0, y, z)$ with $y > 0$, $z > 0$, is
$$G_2 = \left\{ \begin{pmatrix} a & & & \\ & \varepsilon a & & \\ & & \varepsilon a & \\ & & & a \end{pmatrix} : a, \varepsilon \in U(1),\ a^4 = 1,\ \varepsilon^2 = 1 \right\},$$
an eight-element group. Thus there are eight orientation classes of second order frames on f. Its Lie algebra is $\mathfrak{g}_2 = (0)$, so $\mathfrak{m}_2 = \mathfrak{g}_1$ and
$$\Theta_2 = \omega_1 = \frac{1}{2}\begin{pmatrix} \omega_1^1-\omega_4^4 & & & \\ & \omega_2^2-\omega_3^3 & & \\ & & \omega_3^3-\omega_2^2 & \\ & & & \omega_4^4-\omega_1^1 \end{pmatrix}.$$

If u is any second order frame field along f, we set $u^*\omega_1^3 = \theta$, so $u^*\omega_2^4 = k\theta$, k constant $0 < k < 1$, $u^*\omega_2^3 = 0 = u^*\omega_1^4$, $u^*(\omega_1^1 + \omega_4^4) = 0$, $u^*(\omega_2^1 - k\omega_4^3) = y\theta$ and $u^*(\omega_3^4 - k\omega_1^2) = z\theta$, where $y, z\colon S \to \mathbb{R}^+$ are smooth. Set $u^*(\omega_1^1 - \omega_4^4) = \xi\theta - \bar{\xi}\bar{\theta}$ and $u^*(\omega_2^2 - \omega_3^3) = \eta\theta - \bar{\eta}\bar{\theta}$, where $\xi, \eta\colon S \to \mathbb{C}$ are smooth.

Any other second order frame field along f is given by $\tilde{u} = uK$, where $K = \begin{pmatrix} a & & & \\ & \varepsilon a & & \\ & & \varepsilon a & \\ & & & a \end{pmatrix}$, $a, \varepsilon \in U(1)$, $a^4 = 1 = \varepsilon^2$. Let $\tilde{\theta}$, $\tilde{\xi}$, $\tilde{\eta}$ be the analogs of θ , ξ , η for $\tilde{u}$. Then $\tilde{u}^*(\Theta_0 + \ldots + \Theta_2)$ $= ad(K^{-1})u^*(\Theta_0 + \ldots + \Theta_2) = u^*(\Theta_0 + \ldots + \Theta_2)$ implies $\tilde{\theta} = \theta$, $\tilde{\xi} = \xi$ and $\tilde{\eta} = \eta$. Hence all higher order frames are the same as the second order frames; i.e., $L_j = L_2$, for all $j \geq 2$. The functions ξ and η are third order invariants of f .

The structure equations of f are:

$$d\theta = \frac{1}{2}(\bar{\xi} + \bar{\eta})\theta \wedge \bar{\theta} , \quad k_{;1} = 0 , \quad y^2 + z^2 = (1 - k^2)^2 ,$$

$$y_{;1} = y\xi , \quad z_{;1} = z\eta ,$$

$$2 \operatorname{Re} \xi_{;2} = |\xi|^2 + \operatorname{Re} \xi\bar{\eta} + 2z^2(1 + k^2)(1 - k^2)^{-2} ,$$

and

$$2 \operatorname{Re} \eta_{;2} = |\eta|^2 + \operatorname{Re} \eta\bar{\xi} + 2y^2(1 + k^2)(1 - k^2)^{-2} .$$

Computing $0 = ddy$ and $0 = ddz$, it follows that $\operatorname{Im} \xi_{;1}$ $= \frac{1}{2} \operatorname{Im} \xi\bar{\eta}$ and $\operatorname{Im} \eta_{;2} = \frac{1}{2} \operatorname{Im} \eta\bar{\xi}$. Thus Re may be omitted from the last two structure equations. Notice that y is nonconstant.

There are two possibilities in the fine structure:

<u>Type IcIIci</u>: y and ξ are independent, the Frenet frames have order 3;

<u>Type IcIIcii</u>: all invariant are functions of y , the Frenet frames

have order 2. There is a 1-parameter subgroup H of G acting non-transitively on $f(S)$ by translation along the y = constant curves in S .

<u>Type IcIId</u>.

We say a Type Ic curve $f: S \to M$ is of Type IcIId if for some, hence any, first order frame field u along f , $u^*(\omega_1^1 + \omega_4^4) \neq 0$ and either $u^*(\omega_2^1 - k\omega_4^3) \neq 0$ or $u^*(\omega_3^4 - k\omega_1^2) \neq 0$; i.e., $x \neq 0$ and either $y \neq 0$ or $z \neq 0$. To be specific we suppose $y \neq 0$. Then there exists, at least locally, a first order frame field along f for which $x > 0$ and $y > 0$. This condition characterizes second order frame fields along f , and the positive smooth functions x and y together with the complex valued function z are second order invariants of f .

The isotropy subgroup of G_1 at any point (x, y, z) with $x > 0$, $y > 0$, is $G_2 = \{aI_4 ; a^4 = 1\}$, a four element group. Thus there are four orientation classes of second order frames on f .

Let u be a second order frame field along f and set $u^*\omega_1^3 = \theta$. As for Type IcIIc curves, θ is independent of the choice of second order frame field along f . Setting $u^*(\omega_1^1 - \omega_4^4) = \xi\theta - \bar{\xi}\bar{\theta}$ and $u^*(\omega_2^2 - \omega_3^3) = \eta\theta - \bar{\eta}\bar{\theta}$, then the smooth functions $\xi, \eta: S \to \mathbb{C}$ are third order invariants of f . Furthermore, $L_j = L_2$ for all $j \geq 2$.

The structure equations of f are:

$$d\theta = (x + \frac{1}{2}(\bar{\xi} + \bar{\eta}))\theta \wedge \bar{\theta} , \qquad k_{;1} = 2kx ,$$

$$2 \operatorname{Re} x_{;1} = 2x^2 + x \operatorname{Re}(\xi + \eta) + 1 - k^2 - (1 - k^2)^{-1}(y^2 + |z|^2) ,$$

$$y_{;1} = y(\xi + 2x) - 4kxz(1 - k^2)^{-1} ,$$

$$z_{;2} = z(\bar{\eta} + 2x) - 4kxy(1 - k^2)^{-1} ,$$

$$2 \operatorname{Re} \xi_{;2} = 2x \operatorname{Re} \xi + |\xi|^2 + \operatorname{Re}(\xi\bar{\eta}) + (1 + k^2)((1 - k^2)^{-2}(|z|^2 - y^2) + 1) ,$$

$$2 \operatorname{Re} \eta_{;2} = 2x \operatorname{Re} \eta + |\eta|^2 + \operatorname{Re}(\xi\bar{\eta}) + (1 + k^2)((1 - k^2)^{-2}(y^2 - |z|^2) + 1) .$$

We omit discussion of the fine structure of Type IcIId.

This completes the Frenet frame construction on holomorphic curves in $\mathbb{C}G_{4,2}$

VI. <u>Special affine surface theory</u>.

In this example we analyze imbedded surfaces in $\mathbb{R}^3$ under the action of the special affine group. Contrasting this example with the example of Chapter II illustrates how the differential geometry of submanifolds depends on the group of transformations under consideration.

This analysis leads to four types at the second order: Types IIa (planar), IIb (parabolic), IIc (elliptic), and IId (hyperbolic). We go on to analyze the first three of these, but omit further analysis of the fourth. The resulting types are:

IIa - (open submanifold of a plane);

IIb - IIIaIva (cylinder), IIIaIVb (cylinder), IIIbVaVIa (cone), IIIbVb;

IIc - IIIa (quadric surfaces not cones or cylinders), IIIbIVa (group manifold).

We call attention to two cases which illustrate features of the frame construction which do not arise for any of the other examples. In the analysis of the Type IIbIIIb surface at the fourth order, it unavoidably occurs that the isotropy subgroup of (G_3, ρ_3) depends on the point in the cross section. And in the analysis of the Type IIbIIIbVa, the covariant derivative of the fifth order invariant plays the essential role in determining the sixth order frames.

As our primary purpose is to illustrate the method of the frame construction, we refer to Guggenheimer [G, 1963], pp. 290-313, for the details of the many geometric interpretations.

Now $M = \mathbb{R}^3$, and the group G is the semi-direct product $SL(3; \mathbb{R}) \cdot \mathbb{R}^3$, which acts on M by $(A, x)y = Ay + x$, where $A \in SL(3; \mathbb{R})$, $x, y \in \mathbb{R}^3$. The composition law in G is given explicitly by $(A, x)(B, y) = (AB, Ay + x)$, where $A, B \in SL(3; \mathbb{R})$, $x, y \in \mathbb{R}^3$.

The Lie algebra of G is $\mathfrak{g} = \mathfrak{sl}(3; \mathbb{R}) + \mathbb{R}^3$, whose Lie bracket operation is given explicitly by $[(X, x) , (Y, y)] = ([X, Y] , Xy - Yx)$, for $X, Y \in \mathfrak{sl}(3; \mathbb{R})$, $x, y \in \mathbb{R}^3$. For future reference, the inverse of $(A, x) \in G$ is $(A, x)^{-1} = (A^{-1}, -A^{-1}x)$, and the adjoint action of $G_0 = SL(3; \mathbb{R})$ on $\mathfrak{g}$ is $Ad(A)(X, x) = (AXA^{-1}, Ax)$. The exponential map $\exp: \mathfrak{g} \to G$ is given by $\exp(X, x) = (e^X, \frac{e^X - 1}{X} \cdot x)$, where $e^X = \sum_0^\infty \frac{X^j}{j!}$ and $\frac{e^X - 1}{X} = \sum_1^\infty \frac{X^{j-1}}{j!}$.

The Maurer-Cartan form of G is $\Omega_{(A,x)} = (A^{-1}dA, A^{-1}dx) = ((\omega^i_j), (\theta^i))$, where ω^i_j , θ^i are left-invariant 1-forms on G , $\sum_1^3 \omega^i_i = 0$, and $1 \leq i , j \leq 3$. The Maurer-Cartan equations $d\Omega = -\frac{1}{2}[\Omega, \Omega]$ of G are $d\omega^i_j = -\sum_{k=1}^3 \omega^i_k \wedge \omega^k_j$ and $d\theta^i = -\sum_{k=1}^3 \omega^i_k \wedge \theta^k$, $1 \leq i , j \leq 3$.

The isotropy subgroup of G at the origin 0 of $\mathbb{R}^3$ is $G_0 = SL(3; \mathbb{R})$, and its Lie algebra is $\mathfrak{g}_0 = \mathfrak{sl}(3; \mathbb{R})$. Let $\pi: G \to G/G_0 = M$ denote the natural projection. In contrast to the previous examples, the isotropy subgroup is non-compact.

As a vector subspace of $\mathfrak{g}$ complementary to $\mathfrak{g}_0$ we take $\mathfrak{m}_0 = \mathbb{R}^3$, with the standard basis. The decomposition $\mathfrak{g} = \mathfrak{g}_0 + \mathfrak{m}_0$ decomposes Ω into $\Omega = \omega_0 + \Theta_0$, where $\omega_0 = (\omega^i_j)$ is the $\mathfrak{g}_0$-component and $\Theta_0 = (\theta^i)$ is the $\mathfrak{m}_0$-component.

We consider a smooth imbedding $f: S \to M$, where S is a connected 2-manifold. The zeroth order frames on f are the elements of $L_0 = \{(A, f(s)): A \in SL(3; \mathbb{R}), s \in S\} \subseteq G$. A zeroth order frame field along f is a smooth map $u: S \to G$ such that $\pi \circ u = f$; i.e., $u = (A, f)$, where $A: S \to SL(3; \mathbb{R})$ is smooth.

To construct the first order frames, let ϕ^1, ϕ^2 be a coframe field on S, i.e., a pair of everywhere linearly independent 1-forms on S. Let u be a zeroth order frame field along f and set $u^*\theta^i = \sum_{\alpha=1}^{2} x^i_\alpha \phi^\alpha$, where x^i_α is a smooth function on S, $1 \leqslant i \leqslant 3$, $1 \leqslant \alpha \leqslant 2$. Then $\lambda_0 \circ u: S \to G_{3,2}$ is given by
$$\lambda_0 \circ u = \begin{bmatrix} x^1_1 & x^1_2 \\ x^2_1 & x^2_2 \\ x^3_1 & x^3_2 \end{bmatrix}.$$

Any other zeroth order frame field along f is given by $\tilde{u} = uK$, where $K = (A, 0)$, $A: S \to SL(3; \mathbb{R})$. Letting $\tilde{x}^i_\alpha$ be the analog of x^i_α for $\tilde{u}$, we have, by the $ad(G_0)$-invariance of $\mathfrak{m}_0$, that $\tilde{u}^*\Theta_0 = ad(K^{-1})u^*\Theta_0$, and hence $\lambda_0 \circ \tilde{u} = A \cdot \lambda_0 \circ u$. Since the standard action of $SL(3; \mathbb{R})$ on $G_{3,2}$ is transitive, there exists, at least locally, a zeroth order frame field u along f such that
$$\lambda_0 \circ u = \begin{bmatrix} 1 & 0 \\ 0 & 1 \\ 0 & 0 \end{bmatrix};$$
i.e., such that $u^*\theta^3 = 0$ and $u^*\theta^1$, $u^*\theta^2$ is a coframe field on S.

This condition characterizes the first order frame fields along f . There are no first order invariants.

The isotropy subgroup of G_0 at $\begin{bmatrix} 1 & 0 \\ 0 & 1 \\ 0 & 0 \end{bmatrix}$ is $G_1 = \{A \in G_0 : A_{31} = 0 = A_{32}\}$, where A_{ij} denotes the ij^{th} entry of $A \in SL(3;\mathbb{R}) = G_0$. Its Lie algebra is $\mathfrak{g}_1 = \{X \in \mathfrak{g}_0 : X_{31} = 0 = X_{32}\}$. For a subspace of $\mathfrak{g}_0$ complementary to $\mathfrak{g}_1$ we take $\mathfrak{m}_1$

$$= \left\{ \begin{pmatrix} 0 & & 0 \\ & & 0 \\ r & t & 0 \end{pmatrix} \in \mathfrak{g}_0 : r, t \in \mathbb{R} \right\}.$$

Observe that in this case it is impossible to choose $\mathfrak{m}_1$ to be $Ad(G_1)$-invariant. The decomposition $\mathfrak{g}_0 = \mathfrak{g}_1 + \mathfrak{m}_1$ decomposes ω_0 into $\omega_0 = \omega_1 + \Theta_1$, where ω_1 is the $\mathfrak{g}_1$-component and $\Theta_1 = \omega_1^3 \otimes E_4 + \omega_2^3 \otimes E_5$, where $E_4 = \begin{pmatrix} 0 & 0 & 0 \\ 0 & 0 & 0 \\ 1 & 0 & 0 \end{pmatrix}$, $E_5 = \begin{pmatrix} 0 & 0 & 0 \\ 0 & 0 & 0 \\ 0 & 1 & 0 \end{pmatrix}$ is a basis of $\mathfrak{m}_1$.

If u is a first order frame field along f , we let $\phi^i = u^*\theta^i$, $i = 1, 2$ Taking the exterior derivative of $u^*\theta^3 = 0$, and applying the Maurer-Cartan equations of G , it follows that $u^*\omega_1^3 = x\theta^1 + y\phi^2$ and $u^*\omega_2^3 = y\phi^1 + z\phi^2$, for smooth functions x, y, z on S . Let $X = \begin{pmatrix} x & y \\ y & z \end{pmatrix}$.

Any other first order frame field along f is given by $\tilde{u} = uK$, where $K = \begin{pmatrix} A & \xi \\ 0\ 0 & a \end{pmatrix}$, $A: S \to GL(2;\mathbb{R})$, $\xi: S \to \mathbb{R}^2$, all smooth maps, and $a = (\det A)^{-1}$. Let $\tilde{\phi}^1, \tilde{\phi}^2$, and $\tilde{X} = \begin{pmatrix} \tilde{x} & \tilde{y} \\ \tilde{y} & \tilde{z} \end{pmatrix}$ be the analogs of ϕ^1, ϕ^2, X for $\tilde{u}$. Then $\tilde{u}^*(\Theta_0 + \Theta_1) =$ the $(\mathfrak{m}_0 + \mathfrak{m}_1)$-component of $ad(K^{-1})u^*(\Theta_0 + \Theta_1)$ implies that $\begin{pmatrix} \tilde{\phi}^1 \\ \tilde{\phi}^2 \end{pmatrix} = A^{-1}\begin{pmatrix} \phi^1 \\ \phi^2 \end{pmatrix}$ and $\tilde{u}^*(\omega_1^3, \omega_2^3)$ $= a^{-1}u^*(\omega_1^3, \omega_2^3)A$. Combining these we have $\tilde{X} = a^{-1}\,{}^tAXA$.

The action of G_1 on the vector space $\mathcal{S}_2$, of all 2×2 real symmetric matrices, given by $\begin{pmatrix} A & \xi \\ 0\;0 & a \end{pmatrix} X = a\,{}^tA^{-1}XA^{-1}$ has exactly four orbits. Each of the following four matrices represents one of these orbits: $\begin{pmatrix} 0 & 0 \\ 0 & 0 \end{pmatrix}$, Type IIa, (planar surfaces); $\begin{pmatrix} 1 & 0 \\ 0 & 0 \end{pmatrix}$, Type IIb, (parabolic surfaces); $\begin{pmatrix} 1 & 0 \\ 0 & 1 \end{pmatrix}$, Type IIc, (elliptic surfaces); and $\begin{pmatrix} 1 & 0 \\ 0 & -1 \end{pmatrix}$, Type IId, (hyperbolic surfaces). Due to the formal similarities between Types IIc and IId, we shall discuss only the first three types.

The following lemma is elementary.

Lemma. Suppose $X: S \to \mathcal{S}_2$ is a smooth map such that $X(S)$ is contained in a single orbit of G_1 . Then, at least locally, there exists a smooth map $K: S \to G_1$ such that $K(s)$ sends $X(s)$ into the appropriate 1-point cross section of this orbit, for each s in the domain of K .

Type IIa, (planar surfaces).

We say a surface $f: S \to M$ is of Type IIa if for some, hence any, zeroth order frame field u along f we have $u^*\omega_1^3 = 0 = u^*\omega_2^3$; i.e., $X = 0$. Then all higher order frames are the same as the first order frames; i.e., $L_j = L_1$, for all $j \geq 1$. There are no invariants.

By section I.12, $f(S)$ is G-congruent to an open submanifold of $\pi(H)$, where H is the 8-dimensional analytic subgroup of G whose Lie algebra is $\mathfrak{h} = \{\theta^3 = 0 , \omega_1^3 = 0 , \omega_2^3 = 0\}$. Then H is the semi-direct product $G_1 \cdot \mathbb{R}^2$, where $\mathbb{R}^2 \subseteq \mathbb{R}^3$ as $\left\{\begin{pmatrix} x \\ y \\ 0 \end{pmatrix} : x, y \in \mathbb{R}\right\}$, and $\pi(H)$ is the xy-plane in $\mathbb{R}^3$.

Type IIb, (parabolic surfaces).

We say a surface $f: S \to M$ is of Type IIb if, at least locally, there exists a first order frame field u along f such that $u^*\omega_1^3 = \phi^1$ and $u^*\omega_2^3 = 0$; i.e., $X = \begin{pmatrix} 1 & 0 \\ 0 & 0 \end{pmatrix}$. This condition characterizes the second order frame fields along f . There are no second order invariants.

The isotropy subgroup of G_1 at $\begin{pmatrix} 1 & 0 \\ 0 & 0 \end{pmatrix}$ is G_2 $= \left\{\begin{pmatrix} a & 0 & c \\ b & a^{-3} & t \\ 0 & 0 & a^2 \end{pmatrix} : a, b, c, t \in \mathbb{R}\right\}$, $a \neq 0$, which has two connected components, corresponding to two orientation classes of second order frames on f . Its Lie algebra is $\mathfrak{g}_2 = \left\{ \begin{pmatrix} r & 0 & t \\ s & -3r & x \\ 0 & 0 & 2r \end{pmatrix} : r, s, t, x \in \mathbb{R} \right\}$,

and as a subspace of $\mathfrak{g}_1$ complementary to $\mathfrak{g}_2$ we take $\mathfrak{m}_2$

$= \left\{ \begin{pmatrix} 3z & w & 0 \\ 0 & z & 0 \\ 0 & 0 & -4z \end{pmatrix} : z, w \in \mathbb{R} \right\}$. The decomposition $\mathfrak{g}_1 = \mathfrak{g}_2 + \mathfrak{m}_2$ decomposes ω_1 into $\omega_1 = \omega_2 + \Theta_2$, where $\omega_2 = \begin{pmatrix} \mu & 0 & \omega_3^1 \\ \omega_1^2 & -3\mu & \omega_3^2 \\ 0 & 0 & 2\mu \end{pmatrix}$,

$\mu = \frac{1}{10}(\omega_1^1 - 3\omega_2^2)$, is the $\mathfrak{g}_2$-component; and $\Theta_2 = \nu \otimes E_6 + \omega_2^1 \otimes E_7$,

where $\nu = \frac{1}{10}(3\omega_1^1 + \omega_2^2)$, and $E_6 = \begin{pmatrix} 3 & 0 & 0 \\ 0 & 1 & 0 \\ 0 & 0 & -4 \end{pmatrix}$, $E_7 = \begin{pmatrix} 0 & 1 & 0 \\ 0 & 0 & 0 \\ 0 & 0 & 0 \end{pmatrix}$

is a basis of $\mathfrak{m}_2$.

Let u be a second order frame field along f , let $u^*\theta^i = \phi^i$, $i = 1, 2$, and then $u^*\theta^3 = 0$, $u^*\omega_1^3 = \phi^1$, and $u^*\omega_2^3 = 0$. Taking the exterior derivative of $u^*\omega_1^3 = \phi^1$ and $u^*\omega_2^3 = 0$, and applying the Maurer-Cartan equations of G , it follows that $u^*\omega_2^1 = q\phi^1$ and $u^*(3\omega_1^1 + \omega_2^2) = p\phi^1 + q\phi^2$, for some smooth functions p and q on S .

Any other second order frame field along f is given by $\tilde{u} = uK$, where $K = \begin{pmatrix} A & \xi \\ 0\ 0 & a^2 \end{pmatrix}$ $A = \begin{pmatrix} a & 0 \\ b & a^{-3} \end{pmatrix}$. $\xi = \begin{pmatrix} c \\ t \end{pmatrix}$, a, b, c t smooth functions on S , $a \neq 0$. Let $\tilde{\phi}^1$, $\tilde{\phi}^2$, $\tilde{p}$, $\tilde{q}$ be the analogs of ϕ^1 , ϕ^2 , p , q for $\tilde{u}$. Then $\tilde{u}^*(\Theta_0 + \Theta_1 + \Theta_2) = (\mathfrak{m}_0 + \mathfrak{m}_1 + \mathfrak{m}_2)$-component of $\mathrm{ad}(K^{-1})u^*(\Theta_0 + \Theta_1 + \Theta_2)$ implies that: $\tilde{\phi}^1 = a^{-1}\phi^1$,

$\tilde{\phi}^2 = -a^2b\phi^1 + a^3\phi^2$, $\tilde{u}^*\omega_2^1 = a^{-4}u^*\omega_2^1$, and $\tilde{u}^*\nu = u^*(\nu + \frac{1}{5}a^{-1}b\omega_2^1 - \frac{9}{10}a^{-2}c\theta^1)$

Hence

$$\tilde{q} = a^{-3}q \quad \text{and} \quad \tilde{p} = ap + 3bq - 3a^{-1}c .$$

The action of G_2 on $\mathbb{R}^2$ given by $\begin{pmatrix} a & 0 & c \\ b & a^{-3} & t \\ 0 & 0 & a^2 \end{pmatrix}(p, q)$ $= (a^{-1}p - 3a^2bq + 3a^{-2}c , \ a^3q)$ has two orbits: the p-axis $\{q = 0\}$ (Type IIbIIIa), for which the point $(0, 0)$ is a cross section; and the complement $\{q \neq 0\}$ (Type IIbIIIb), for which the point $(0, 1)$ is a cross section.

<u>Type IIbIIIa</u>.

We say a Type IIb surface $f: S \to M$ is of Type IIbIIIa if for some, hence any, second order frame field u along f we have $u^*\omega_2^1 = 0$; i.e., $q = 0$. Then there exists a second order frame field u along f such that $u^*(3\omega_1^1 + \omega_2^2) = 0$; i.e., $p = 0$. This condition characterizes the third order frame fields along f . There are no third order invariants.

The isotropy subgroup of G_2 at $(0, 0)$ is G_3 $= \left\{ \begin{pmatrix} a & 0 & 0 \\ b & a^{-3} & t \\ 0 & 0 & a^2 \end{pmatrix} : a, b, t \in \mathbb{R} , \ a \neq 0 \right\}$, which has two connected components, corresponding to two orientation classes of third order frames on f . Its Lie algebra is $\mathfrak{g}_3 = \left\{\begin{pmatrix} a & 0 & 0 \\ b & -3a & c \\ 0 & 0 & 2a \end{pmatrix} : a, b, c \in \mathbb{R}\right\}$, and for a subspace of $\mathfrak{g}_2$ complementary to $\mathfrak{g}_3$ we take $\mathfrak{m}_3 = \left\{\begin{pmatrix} 0 & 0 & 1 \\ 0 & 0 & 0 \\ 0 & 0 & 0 \end{pmatrix} t : t \in \mathbb{R}\right\}$. The decomposition $\mathfrak{g}_2 = \mathfrak{g}_3 + \mathfrak{m}_3$ decomposes ω_2 into $\omega_3 + \Theta_3$, where $\Theta_3 = \begin{pmatrix} 0 & 0 & \omega_3^1 \\ 0 & 0 & 0 \\ 0 & 0 & 0 \end{pmatrix}$, and the $\mathfrak{g}_3$-component is $\omega_3 = \begin{pmatrix} \mu & 0 & 0 \\ \omega_1^2 & -3\mu & \omega_3^2 \\ 0 & 0 & 2\mu \end{pmatrix}$, $\mu = \frac{1}{10}(\omega_1^1 - 3\omega_2^2)$.

Let u be a third order frame field along f, let $u^*\theta^i = \phi^i$, $i = 1, 2$, and then $u^*\theta^3 = 0$, $u^*\omega_1^3 = \phi^1$, $u^*\omega_2^3 = 0$, $u^*\omega_2^1 = 0$, and $u^*(3\omega_1^1 + \omega_2^2) = 0$. Taking the exterior derivative of the last two equations, and applying the Maurer-Cartan equations of G, it follows that $u^*\omega_3^1 = x\phi^1$ for some smooth function x on S.

Any other third order frame field along f is given by $\tilde{u} = uK$, wehre $K = \begin{pmatrix} a & 0 & 0 \\ b & a^{-3} & c \\ 0 & 0 & a^2 \end{pmatrix}$, $a, b\ c$ smooth functions on S, $a \neq 0$.
Let $\tilde{\phi}^1$, $\tilde{\phi}^2$, $\tilde{x}$ be the analogs of ϕ^1, ϕ^2, x for $\tilde{u}$. Then $\tilde{u}^*(\Theta_0 + \ldots + \Theta_3) = (\mathfrak{m}_0 + \ldots + \mathfrak{m}_3)$-component of $\mathrm{ad}(K^{-1})u^*(\Theta_0 + \ldots + \Theta_3)$ implies that: $\tilde{\phi}^1 = a^{-1}\phi^1$, $\tilde{\phi}^2 = -a^2b\phi^1 + a^3\phi^2$ and $\tilde{u}^*\omega_3^1 = au^*\omega_3^1$.
Hence $\tilde{x} = a^2x$.

The action of G_3 on $\mathbb{R}$ given by $\begin{pmatrix} a & 0 & 0 \\ b & a^{-3} & c \\ 0 & 0 & a^2 \end{pmatrix} x = a^{-2}x$ has two orbit types: the origin, $\{x = 0\}$ (Type IIbIIIaIVa); and its complement $\{x \neq 0\}$ (Type IIbIIIaIVb), for which the set $\{\pm 1\}$ is a cross section.

Type IIbIIIaIVa.

We say a Type IIbIIIa surface $f\colon S \to M$ has Type IIbIIIaIVa if for some, hence any, third order frame field u along f we have $u^*\omega_3^1 = 0$; i.e., $x = 0$. This condition characterizes the fourth order frame fields along f , and thus the fourth, and all higher order frames, are the same as the third order frames on f ; i.e., $L_j = L_3$, for all $j \geq 3$. There are no invariants.

It follows from section I.12 that $f(S)$ is G-congruent to an open submanifold of the homogeneous surface $\pi(H)$, where H is the 5-dimensional analytic subgroup of G whose Lie algebra is $\mathfrak{h}$ $= \{\theta^3 = 0 \, , \; \omega_1^3 = \theta^1 \, , \; \omega_2^3 = 0 \, , \; \omega_2^1 = 0 \, , \; 3\omega_1^1 + \omega_2^2 = 0 \, , \; \omega_3^1 = 0\}$. It is easily verified that $\pi(H)$ is the cylinder $\{z = \frac{1}{2}x^2\}$ in $\mathbb{R}^3$.

Type IIbIIIaIVb.

We say a Type IIbIIIa surface $f\colon S \to M$ has Type IIbIIIaIVb if for any third order frame field u along f we have $u^*\omega_3^1 \neq 0$; i.e., $x \neq 0$. Then there exists a third order frame field u along f such that $u^*\omega_3^1 = \pm\phi^1$; i.e., $x = \pm 1$. This condition characterizes the fourth order frame fields along f . The (constant) fourth order invariant is $\varepsilon = \pm 1$.

The isotropy subgroup of G_3 at either point of $\{\pm 1\}$ is $G_4 = \left\{\begin{pmatrix}1 & 0 & 0\\ b & 1 & c\\ 0 & 0 & 1\end{pmatrix} : b, c \in \mathbb{R}\right\}$, which is connected, corresponding to just one orientation class of fourth order frames on f. Its Lie algebra is $\mathfrak{g}_4 = \left\{\begin{pmatrix}0 & 0 & 0\\ b & 0 & c\\ 0 & 0 & 0\end{pmatrix} : b, c \in \mathbb{R}\right\}$, and for a subspace of $\mathfrak{g}_3$ complementary to $\mathfrak{g}_4$ we take $\mathfrak{m}_4 = \left\{\begin{pmatrix}1 & 0 & 0\\ 0 & -3 & 0\\ 0 & 0 & 2\end{pmatrix} a : a \in \mathbb{R}\right\}$. The decomposition $\mathfrak{g}_3 = \mathfrak{g}_4 + \mathfrak{m}_4$ decomposes ω_3 into $\omega_3 = \omega_4 + \Theta_4$, where the $\mathfrak{g}_4$-component is $\omega_4 = \begin{pmatrix}0 & 0 & 0\\ \omega_1^2 & 0 & \omega_3^2\\ 0 & 0 & 0\end{pmatrix}$, and the $\mathfrak{m}_4$-component is

$$\Theta_4 = \mu \otimes \begin{pmatrix}1 & 0 & 0\\ 0 & -3 & 0\\ 0 & 0 & 2\end{pmatrix}, \quad \mu = \frac{1}{10}(\omega_1^1 - 2\omega_2^2) .$$

Let u be a fourth order frame field along f, and set $u^*\theta^1 = \phi^1$, $u^*\theta^2 = \phi^2$, so that $u^*\theta^3 = 0$, $u^*\omega_1^3 = \phi^1$, $u^*\omega_2^3 = 0$, $u^*\omega_2^1 = 0$, $u^*(3\omega_1^1 + \omega_2^2) = 0$, and $u^*\omega_3^1 = \varepsilon\phi^1$. Taking the exterior derivative of the last equation, and applying the Maurer-Cartan equations of G, it follows that $u^*\omega_1^1 = y\phi^1$, for some smooth function y on S.

Any other fourth order frame field along f is given by $\tilde{u} = uK$, where $K = \begin{pmatrix}1 & 0 & 0\\ b & 1 & c\\ 0 & 0 & 1\end{pmatrix}$, b and c smooth functions on S. Let

$\tilde{\phi}^1$, $\tilde{\phi}^2$, $\tilde{y}$ be the analogs of ϕ^1 , ϕ^2 , y for $\tilde{u}$. Then $\tilde{u}^*(\Theta_0 + \ldots + \Theta_4) = (\mathfrak{M}_0 + \ldots + \mathfrak{M}_4)$-component of $\mathrm{ad}(K^{-1})u^*(\Theta_0 + \ldots + \Theta_4)$ implies that: $\tilde{\phi}^1 = \phi^1$, $\tilde{\phi}^2 = -b\phi^1 + \phi^2$, and $\tilde{u}^*\omega_1^1 = u^*\omega_1^1$. Hence ϕ^1 and $u^*\omega_1^1$ are fourth order invariant forms, and $\tilde{y} = y$.

Thus the fifth order frames are the same as the fourth order frames. Furthermore, taking the exterior derivative of $u^*\omega_1^1 = y\phi^1$, and applying the Maurer-Cartan equations of G , it follows that $dy = y_{;1}\phi^1$, (i.e., $y_{;2} = 0$) . Since also $dy = \tilde{y}_{;1}\tilde{\phi}^1 = \tilde{y}_{;1}\tilde{\phi}^1$, it follows that $y_{;1} = \tilde{y}_{;1}$, and hence all higher order frames are the same as the fourth order frames; i.e., $L_j = L_4$, for all $j \geqslant 4$. The function y is the fifth order invariant of f .

The structure equations of f are: $d\phi^1 = 0$ and $dy = y_{;1}\phi^1$. The fine structure is simple, but quite interesting, in this case. Namely, y is either constant (Type IIbIIIaIVbi), or non-constant (Type IIbIIIaIVbii).

Type IIbIIIaIVbi.

Now all invariants of f are constant and we know $f(S)$ must be a piece of a homogeneous surface. In fact, let $\mathcal{G}_{y\varepsilon}$ denote the 4-dimensional, left-invariant, involutive distribution defined on G by the equations $\{\theta^3 = 0$, $\omega_1^3 = \theta^1$, $\omega_2^3 = 0$, $\omega_2^1 = 0$, $\omega_3^1 = \varepsilon\theta^1$, $\omega_1^1 = y\theta^1$, $\omega_2^2 = -3y\theta^1\}$, where $\varepsilon = \pm 1$ is the fourth order invariant of f . Then $\mathfrak{h}_{y\varepsilon} = \left\{ \left(\begin{pmatrix} yp & 0 & \varepsilon p \\ r & -3yp & s \\ p & 0 & 2yp \end{pmatrix}, \begin{pmatrix} p \\ q \\ 0 \end{pmatrix} \right) : p, q, r, s \in \mathbb{R} \right\}$.

Let $H_{y\varepsilon}$ denote the analytic subgroup of G whose Lie algebra is $\mathfrak{h}_{y\varepsilon}$. It is easily checked that $\pi H_{y\varepsilon}$ is a homogeneous Type IIbIIIaIVbi surface with fourth order invariant equal to ε and constant fifth order invariant equal to y . Although it is easily seen that $\pi H_{y\varepsilon}$ is a cylinder with generator parallel to ε_2 , it is difficult to obtain a closed expression for the profile curve, except for special values of y . For example, when $y = 0$, the profile curve in the x^1x^3-plane is: the branch of the hyperbola $(x^3)^2 - (x^1)^2 = 1$, $x^3 > 0$, if $\varepsilon = +1$; and the circle $(x^1)^2 + (x^3 + 1)^2 = 1$, if $\varepsilon = -1$.

<u>Theorem</u>. Let $f: S \to \mathbb{R}^3$ be a Type IIbIIIaIVbi surface with constant fourth and fifth order invariants ε and y , respectively. Then f is G-congruent to an open submanifold of the cylinder $\pi H_{y\varepsilon}$.

<u>Proof</u>: Let $u: S \to G$ be a Frenet frame along f . Then $u(S)$ is an integrable submanifold of $\mathfrak{h}$, so there exists $A \in G$ such that $Au(S) \subset H_{y\varepsilon}$. Replacing f by $A \circ f$, we may assume $u(S) \subset H_{y\varepsilon}$. Hence $f(S) = \pi u(S) \subset \pi H_{y\varepsilon}$.

<u>Type IIbIIIaIVbii</u>. Now the fifth order invariant y of f is nonconstant. Recall that if u is a fourth order frame field along f , and if we set $u^*\theta^1 = \phi^1$, $u^*\theta^2 = \phi^2$, then $u^*\theta^3 = 0$, $u^*\omega_1^3 = \phi^1$, $u^*\omega_2^3 = 0$, $u^*\omega_2^1 = 0$, $u^*(3\omega_1^1 + \omega_2^2) = 0$, $u^*\omega_3^1 = \varepsilon\phi^1$ and $u^*\omega_1^1 = y\phi^1$,

where $\varepsilon = \pm 1$ is the fourth order invariant.

We shall apply Theorem 5 of section I.12 quite explicitly to this example. Let u be a fourth order frame field along f and fix a point $s_0 \in S$ where $dy \neq 0$. Translating f by an element of G, if necessary, we may assume that $f(s_0) = 0$ and $u(s_0) = 1$. Let $y_0 = y(s_0)$ and let $S(y_0) = y^{-1}\{s_0\}$, a curve in S. On $S(y_0)$ we have $0 = dy = y_{;1}\phi^1$, so $\phi^1 = 0$ since $y_{;1} \neq 0$. (y_0 is a regular value of y). Let $\mathfrak{h}$ be the 3-dimensional Lie subalgebra of $\mathfrak{g}$ defined by $\mathfrak{h} = \{\theta^3 = 0\ ,\ \omega_1^3 = \theta^1\ ,\ \omega_2^3 = 0\ ,\ \omega_2^1 = 0\ ,\ 3\omega_1^1 + \omega_2^2 = 0\ ,$

$$\omega_3^1 = \pm\theta^1\ ,\ \omega_1^1 = y_0\theta^1\ ,\ \theta^1 = 0\} = \left\{\left(\begin{pmatrix}0 & 0 & 0\\ r & 0 & s\\ 0 & 0 & 0\end{pmatrix},\ \begin{pmatrix}0\\ t\\ 0\end{pmatrix}\right) : r, s, t \in \mathbb{R}\right\}.$$

Its orbits are lines papallel to ε_2. Hence $f(S)$ is G-congruent to an open submanifold of a cylinder generated by straight lines parallel to ε_2 translated along a profile curve in the $\varepsilon_1\varepsilon_3$-plane.

We point out that the structure equations $d\phi^1 = 0$, $dy = y_{;1}\phi^1$ are the only conditions on ϕ^1 and y. Namely, if S is a domain in $\mathbb{R}^2$ with standard coordinates x, y, and if $\psi(y)$ is any smooth non-vanishing function of y, then for either choice of $\varepsilon = \pm 1$, there exists a smooth imbedding $f: S \to \mathbb{R}^3$ which is a Type IIbIIIaIVbii surface whose invariants (with respect to some fourth order frame field u) are $\phi^1 = \frac{dy}{\psi(y)}$, y and ε.

This concludes the Type IIbIIIa surfaces.

Type IIbIIIb.

We say a Type IIb surface $f: S \to M$ is of Type IIbIIIb if for any second order frame field u along f we have $u^*\omega_2^1 \neq 0$; i.e., $q \neq 0$. Then there exists a second order frame field u along f such that $p = 0$ and $q = 1$; i.e., $u^*\omega_2^1 = \phi^1$ and $u^*(3\omega_1^1 + \omega_2^2) = \phi^2$. This condition characterizes the third order frame fields along f . There are no third order invariants.

The isotropy subgroup of G_2 at $(0, 1)$ is G_3

$$= \left\{ \begin{pmatrix} a & 0 & c \\ b & a^{-3} & t \\ 0 & 0 & a^2 \end{pmatrix} \in G_2 : a^3 = 1 , \ b - ca^{-1} = 0 \right\} = \left\{ \begin{pmatrix} 1 & 0 & a \\ a & 1 & b \\ 0 & 0 & 1 \end{pmatrix} : a, b \in \mathbb{R} \right\} .$$

Its Lie algebra is $\mathfrak{g}_3 = \left\{ \begin{pmatrix} 0 & 0 & r \\ r & 0 & t \\ 0 & 0 & 0 \end{pmatrix} : r, t \in \mathbb{R} \right\}$, and for a subspace of $\mathfrak{g}_2$ complementary to $\mathfrak{g}_3$ we take $\mathfrak{m}_3 = \left\{ \begin{pmatrix} t & 0 & -r \\ r & -3t & 0 \\ 0 & 0 & 2t \end{pmatrix} : t, r \in \mathbb{R} \right\}$.

Then ω_2 , the $\mathfrak{g}_2$-component of Ω , decomposes into $\omega_2 = \omega_3 + \Theta_3$, where the $\mathfrak{m}_3$-component $\Theta_3 = \begin{pmatrix} \mu & 0 & -\sigma \\ \sigma & -3\mu & 0 \\ 0 & 0 & 2\mu \end{pmatrix}$, $\mu = \frac{1}{10}(\omega_1^1 - 3\omega_2^2)$ and $\sigma = \frac{1}{2}(\omega_1^2 - \omega_3^1)$; and the $\mathfrak{g}_3$-component is $\omega_3 = \begin{pmatrix} 0 & 0 & \tau \\ \tau & 0 & \omega_3^2 \\ 0 & 0 & 0 \end{pmatrix}$, $\tau = \frac{1}{2}(\omega_1^2 + \omega_3^1)$.

Let u be a third order frame field along f , and let $u^*\theta^i = \phi^i$, $i = 1, 2$. Then $u^*\theta^3 = 0$, $u^*\omega_1^3 = \phi^1$, $u^*\omega_2^3 = 0$, $u^*\omega_2^1 = \phi^1$, and $u^*(3\omega_1^1 + \omega_2^2) = \phi^2$. Taking the exterior derivative of the last two equations, and applying the Maurer-Cartan equations of G , it follows that $u^*\omega_1^1 = -m\phi^1$, $u^*\omega_2^2 = 3m\phi^1 + \phi^2$, and $u_1^*(\omega_1^2 - \omega_3^1) = n\phi^1 + m\phi^2$, for some smooth functions m and n on S .

Any other third order frame field along f is given by $\tilde{u} = uK$, where $K = \begin{pmatrix} 1 & 0 & a \\ a & 1 & b \\ 0 & 0 & 1 \end{pmatrix}$, a and b smooth functions on S . Let $\tilde{\phi}^1$, $\tilde{\phi}^2$, $\tilde{m}$, $\tilde{n}$ be the analogs of ϕ^1 , ϕ^2 , m , n for $\tilde{u}$. Then $\tilde{u}^*(\Theta_0 + \ldots + \Theta_3) = (\mathfrak{m}_0 + \ldots + \mathfrak{m}_3)$-component of $\mathrm{ad}(K^{-1})u^*(\Theta_0 + \ldots + \Theta_3)$ implies that $\tilde{\phi}^1 = \phi^1$, $\tilde{\phi}^2 = \phi^2 - a\phi^1$, $\tilde{u}^*\omega_1^1 = u^*\omega_1^1$, $\tilde{u}^*\omega_2^2 = u^*\omega_2^2 - a\phi^1$, and $\tilde{u}^*\sigma = u^*\sigma - \frac{3}{2}au^*\omega_1^1 + \frac{1}{2}(a^2 - 2b)\phi^1$. Hence

$$\tilde{m} = m \quad \text{and} \quad \tilde{n} = n + 4am + a^2 - 2b \ .$$

Note that ϕ^1 and $u^*\omega_1^1$ are third order invariant forms on S , i.e., they are independent of the choice of third order frame field along f . It is evident that we may choose a and b so that $n = 0$. Thus we define a <u>fourth order frame</u> field along f to be a third order frame field u for which $n = 0$, i.e., $u^*(\omega_1^2 - \omega_3^1) = m\phi^2$. The smooth function m is a <u>fourth order invariant</u> of f .

The isotropy subgroup of G_3 at $(m, 0)$ is G_4

$$= \left\{ \begin{pmatrix} 1 & 0 & a \\ a & 1 & b \\ 0 & 0 & 1 \end{pmatrix} : 4am + a^2 - 2b = 0 \right\} = \left\{ \begin{pmatrix} 1 & 0 & a \\ a & 1 & 2am+\frac{a^2}{2} \\ 0 & 0 & 0 \end{pmatrix} : a \in \mathbb{R} \right\} .$$

This is the first case among our examples in which the isotropy subgroup depends on the point in the particular local cross section! That is, G_4

is a 1-parameter family of subgroups, $G_4(m)$, depending on m . This dependence is unavoidable; that is, there is no way to choose a cross section of the action of G_3 so that the isotropy subgroup of G_3 is independent of the point in the cross section.

The Lie algebra of G_4 is $\mathfrak{g}_4 = \left\{\begin{pmatrix} 0 & 0 & a \\ a & 0 & 2ma \\ 0 & 0 & 0 \end{pmatrix} : a \in \mathbb{R}\right\}$, and as a complementary subspace of $\mathfrak{g}_4$ in $\mathfrak{g}_3$ we take $\mathfrak{m}_4 = \left\{\begin{pmatrix} 0 & 0 & 0 \\ 0 & 0 & r \\ 0 & 0 & 0 \end{pmatrix} : r \in \mathbb{R}\right\}$. Then ω_3 , the $\mathfrak{g}_3$-component of Ω , decomposes into $\omega_3 = \omega_4 + \Theta_4$, where the $\mathfrak{m}_4$-component is $\Theta_4 = \begin{pmatrix} 0 & 0 & 0 \\ 0 & 0 & \omega_3^2 - 2m\tau \\ 0 & 0 & 0 \end{pmatrix}$, $\tau = \frac{1}{2}(\omega_1^2 + \omega_3^1)$, and the $\mathfrak{g}_4$-component is

$$\omega_4 = \begin{pmatrix} 0 & 0 & \tau \\ \tau & 0 & 2m\tau \\ 0 & 0 & 0 \end{pmatrix} .$$

Let u be a fourth order frame field along f , and let $u^*\theta^i = \phi^i$, $i = 1, 2$. Then $u^*\theta^3 = 0$, $u^*\omega_1^3 = \phi^1$, $u^*\omega_2^3 = 0$, $u^*\omega_2^1 = \phi^1$, $u^*\omega_1^1 = -m\phi^1$, $u^*\omega_2^2 = 3m\phi^1 + \phi^2$, and $u^*(\omega_1^2 - \omega_3^1) = m\phi^2$. Taking the exterior derivative of the last two equations, and applying the Maurer-Cartan equations of G , it follows that $m_{;2} = -2m$ and $m_{;1} = 2t$, where $u^*(\omega_3^2 - m(\omega_1^2 + \omega_3^1)) = r\phi^1 + t\phi^2$ for some smooth functions r and t on S , and $dm = m_{;1}\phi^1 + m_{;2}\phi^2$.

Any other fourth order frame field along f is given by $\tilde{u} = uK$,

where $K = \begin{pmatrix} 1 & 0 & a \\ a & 1 & 2ma+\frac{a^2}{2} \\ 0 & 0 & 1 \end{pmatrix}$, where a is any smooth function on

S. Let $\tilde{\phi}^1, \tilde{\phi}^2, \tilde{r}, \tilde{t}$ be the analogs of ϕ^1, ϕ^2, r, t for $\tilde{u}$. Then $\tilde{u}^*(\Theta_0 + \ldots + \Theta_4 = (\mathfrak{M}_0 + \ldots + \mathfrak{M}_4)$-component of $ad(K^{-1})u^*(\Theta_0 + \ldots + \Theta_4)$ implies that: $\tilde{\phi}^1 = \phi^1$, $\tilde{\phi}^2 = \phi^2 - a\phi^1$, $\tilde{u}^*\omega_2^2 = u^*\omega_2^2 - a\phi^1$, and $\tilde{u}^*(\omega_3^2 - 2m\tau) = u^*\{\omega_3^2 - 2m\tau + 2a\sigma - 3a^2m\theta^1 + (\frac{-3}{2}a^2 + am)\omega_1^1 + 2am\omega_2^2\}$, (where $\sigma = \frac{1}{2}(\omega_1^2 - \omega_3^1)$ and $\tau = \frac{1}{2}(\omega_1^2 + \omega_3^1)$). Hence

$$\tilde{t} = t + 3am \quad \text{and} \quad \tilde{r} = r + at + 5am^2 + \frac{3}{2}a^2m .$$

The type of the action of $G_4(m)$ on $\mathbb{R}^2$ given by $(a, m)\cdot(r, t) = (r - at + \frac{3}{2}a^2m - 5am^2, t - 3am)$, $a, r, t \in \mathbb{R}$. depends on whether m is 0 or not. Now $m = 0$ iff m is constant, since $m_{;2} = -2m$; and if m is constant then $t = 0$, since $m_{;1} = 2t$. There are two types: $m = 0$, and then a cross section of the action is $W_5 = \{(r, 0): r \in \mathbb{R}\}$, (Type IIbIIIbVa); and $m \neq 0$, in which case a cross section is again W_5, (Type IIbIIIbVb). The two cases are separated, of course, because the isotropy subgroups at points of W_5 are different, i.e., have different dimensions, depending on whether m is 0 or not.

Type IIbIIIbVa.

We say a Type IIbIIIb surface $f: S \to M$ is of Type IIbIIIbVa if $m = 0$. Thus $t = 0$ and r is independent of the choice of fourth order frame field. Consequently the fifth order frames are the same as the fourth order frames; i.e., $L_5 = L_4$. The smooth function r on S is the fifth order invariant of f.

The isotropy subgroup of $G_4(0)$ at any point of W_5 is $G_5 = G_4(0)$. Its Lie algebra is $\mathfrak{g}_5 = \mathfrak{g}_4(0)$, and for a subspace of $\mathfrak{g}_4(0)$ complementary to $\mathfrak{g}_5$ we must take $\mathfrak{m}_5 = (0)$. Then $\omega_4 = \omega_5 + \Theta_5$ where $\Theta_5 = 0$.

Let u be a fifth order frame field along f, and let $u^*\theta^i = \phi^i$, $i = 1, 2$. Then $u^*\theta^3 = 0$, $u^*\omega_1^3 = \phi^1$, $u^*\omega_2^3 = 0$, $u^*\omega_2^1 = \phi^1$, $u^*\omega_1^1 = 0$, $u^*\omega_2^2 = \phi^2$, $u^*\omega_1^2 = u^*\omega_3^1$, $u^*\omega_3^2 = r\phi^1$, and $dr = r_{;1}\phi^1 + r_{;2}\phi^2$. Taking the exterior derivative of $u^*\omega_3^2 = r\phi^1$ and applying the Maurer-Cartan equations of G, it follows that $r_{;2} = -3r$.

Any other fifth order frame field along f is given by $\tilde{u} = uK$, where $K = \begin{pmatrix} 1 & 0 & a \\ a & 1 & \frac{a^2}{2} \\ 0 & 0 & 0 \end{pmatrix}$, $a: S \to \mathbb{R}$. Let $\tilde{\phi}^1, \tilde{\phi}^2, \tilde{r}_{;1}$ be the analogs of $\phi^1, \phi^2, r_{;1}$ for $\tilde{u}$. Then $\tilde{u}^*(\Theta_0 + \dots + \Theta_5) = (\mathfrak{m}_0 + \dots + \mathfrak{m}_5)$-component of $\mathrm{ad}(K^{-1})u^*(\Theta_0 + \dots + \Theta_5)$ implies that: $\tilde{\phi}^1 = \phi^1$ and $\tilde{\phi}^2 = \phi^2 - a\phi^1$. Hence

$$\tilde{r}_{;1} = r_{;1} - 3ar .$$

Thus we must consider the action of G_5 on $\mathbb{R}^2$ given by $a\cdot(r, r_{;1}) = (r, r_{;1} + 3ar)$, $a, r, r_{;1} \in \mathbb{R}$. If $r = 0$, then $r_{;1} = 0$ and we just have trivial action on the origin $(0, 0)$, (Type IIbIIIbVaVIa); if $r \neq 0$, and thus non-constant since $r_{;2} = -3r$, a cross section of the action is $W_6 = \{(r, 0): 0 \neq r \in \mathbb{R}\}$, (Type IIbIIIbVaVIb).

Type IIbIIIbVaVIa.

We say a Type IIbIIIbVa surface $f: S \to M$ is of Type IIbIIIbVaVIa if $r = 0$; i.e., $u^*\omega_3^2 = 0$, for any fourth order frame field u along f . There are no invariants and all higher order frames are the same as the fifth order frames, which are the same as the fourth order frames; i.e., $L_j = L_4$ for all $j \geq 4$. The Corollary of Theorem 5 in section I.12 applies.

Let $\mathfrak{h}$ be the 3-dimensional, left-invariant, involutive distribution defined on G defined by the equations $\{\theta^3 = 0 , \omega_1^3 = \theta^1 , \omega_2^3 = 0 , \omega_1^1 = 0 , \omega_2^2 = \theta^2 , \omega_2^1 = \theta^1 , \omega_1^2 = \omega_3^1 , \omega_3^2 = 0\}$. Then by $\mathfrak{h} = \left\{\left(\begin{pmatrix} 0 & x & z \\ z & y & 0 \\ x & 0 & -y \end{pmatrix}, \begin{pmatrix} x \\ y \\ 0 \end{pmatrix}\right) : x, y, z \in \mathbb{R}\right\}$. Let H denote the analytic subgroup of G whose Lie algebra is $\mathfrak{h}$.

Theorem. Any Type IIbIIIbVaVIa surface is G-congruent to an open submanifold of the half-cone C_0 in $\mathbb{R}^3$ defined by $x^2 + y^2 = z^2$, $z > 0$. This half-cone is in fact a homogeneous Type IIbIIIbVaVIa surface.

<u>Proof</u>: Let $u: S \to G$ be a fourth order frame field along f. Then $u(S)$ is an integral submanifold of the distribution $\mathfrak{h}$. Hence a G-translate of $u(S)$ is contained in H. Translating f by the same amount, we may as well assume that $u(S) \subset H$.

In order to determine $\pi(H)$, we exponentiate some members of

$$\mathfrak{h}, \text{ say } \exp\left(\begin{pmatrix} 0 & t & 0 \\ 0 & 0 & 0 \\ t & 0 & 0 \end{pmatrix}, \begin{pmatrix} t \\ 0 \\ 0 \end{pmatrix}\right) = \left(\begin{pmatrix} 1 & t & 0 \\ 0 & 1 & 0 \\ t & \frac{t^2}{2} & 1 \end{pmatrix}, \begin{pmatrix} t \\ 0 \\ \frac{1}{2}t^2 \end{pmatrix}\right) \text{ and}$$

$$\exp\left(\begin{pmatrix} 0 & 0 & 0 \\ 0 & r & 0 \\ 0 & 0 & -r \end{pmatrix}, \begin{pmatrix} 0 \\ r \\ 0 \end{pmatrix}\right) = \left(\begin{pmatrix} 1 & 0 & 0 \\ 0 & e^r & 0 \\ 0 & 0 & e^{-r} \end{pmatrix}, \begin{pmatrix} 0 \\ e^r-1 \\ 0 \end{pmatrix}\right).$$

It is easily determined then that $\pi(H)$ is the cone C defined by $x^2 = 2z(y+1)$, $y > -1$, $z > 0$. Now $TC = C_0$, where $T = (A, a) \in G$,

$$A = \frac{1}{\sqrt{2}}\begin{pmatrix} -\sqrt{2} & 0 & 0 \\ 0 & -1 & 1 \\ 0 & 1 & 1 \end{pmatrix}, \text{ and } a = \frac{1}{\sqrt{2}}\begin{pmatrix} 0 \\ -1 \\ 1 \end{pmatrix}.$$

The group acting on C_0 is then THT^{-1}, whose Lie algebra is

$$Ad(T)\mathfrak{h} = \left\{\left(\begin{pmatrix} 0 & p & q \\ -p & 0 & r \\ q & r & 0 \end{pmatrix}, 0\right) : p, q, r \in \mathbb{R}\right\}.$$

Hence $f(S) = \pi u(S) \subset \pi(H) = C = T^{-1}C_0$, i.e., $Tf(S) \subset C_0$.

In order to show that $C_0 = \{x^2 + y^2 = z^2,\ z > 0\} \subseteq \mathbb{R}^3$ is a Type IIbIIIbVaVIa surface, we construct a zeroth order frame field v on the neighborhood $\left\{\begin{pmatrix} x \\ y \\ z \end{pmatrix} \in C_0 : -\pi < y < \pi\right\}$ by:

$$v\begin{pmatrix}x\\y\\z\end{pmatrix} = \begin{pmatrix}\cosh x\cos y & \cosh x\sin y & \sinh x\\ -\sin y & \cos y & 0\\ \sinh x\cos y & \sinh x\sin y & \cosh x\end{pmatrix}$$

$$= \exp\left(\begin{pmatrix}0&0&x\\0&0&0\\x&0&0\end{pmatrix}, 0\right)\cdot\exp\left(\begin{pmatrix}0&y&0\\-y&0&0\\0&0&0\end{pmatrix}, 0\right),$$ which is actually in

THT^{-1} , since $\left(\begin{pmatrix}0&0&x\\0&0&0\\x&0&0\end{pmatrix}, 0\right)$, $\left(\begin{pmatrix}0&y&0\\-y&0&0\\0&0&0\end{pmatrix}, 0\right) \in Ad(T)\mathfrak{h}$. Then

$u = T^{-1}vT = ad(T^{-1})\circ v$ is a zeroth order frame field on C_0 whose image lies in H . Hence $u^*\Omega$ is an $\mathfrak{h}$-valued 1-form on C_0 , which implies that C_0 is a Type IIbIIIbVaVIa surface.

Type IIbIIIbVaVIb.

We say a Type IIbIIIbVa surface $f: S \to M$ is of Type IIbIIIbVaVIb if $r \neq 0$; i.e., $u^*\omega_3^2 \neq 0$ for any fifth order frame field u along f . Then there exists a fifth order frame field u such that $r_{;1} = 0$. This condition characterizes the sixth order frame fields along f . There are no sixth order invariants (because $r_{;2} = -3r$ is a fifth order invariant).

Now $G_6 = \left\{\begin{pmatrix}1&0&a\\a&1&\frac{a^2}{2}\\0&0&1\end{pmatrix} : ar = 0\right\} = \{1\}$. Hence $L_j = L_6$, for all $j \geq 6$. Since $\mathfrak{g}_6 = (0)$, we have $\mathfrak{m}_6 = \mathfrak{g}_5 = \left\{\begin{pmatrix}0&0&a\\a&0&0\\0&0&0\end{pmatrix} : a \in \mathbb{R}\right\}$,

and $\Theta_6 = \omega_5 = \omega_4 = \tau\begin{pmatrix}0&0&1\\1&0&0\\0&0&0\end{pmatrix}$, $\tau = \frac{1}{2}(\omega_1^2 + \omega_3^1)$.

Let u be a sixth order frame field along f, and let $u^*\phi^i = \phi^i$, $i = 1, 2$. Then $u^*\phi^3 = 0$, $u^*\omega_1^3 = \phi^1$, $u^*\omega_2^3 = 0$, $u^*\omega_2^1 = \phi^1$, $u^*\omega_1^1 = 0$, $u^*\omega_2^2 = \phi^2$, $u^*\omega_1^2 = u^*\omega_3^1$, $u^*\omega_3^2 = r\phi^1$, and $dr = -3r\phi^2$. Taking the exterior derivative of this last equation and applying the Maurer-Cartan equations, it follows that $u^*(\omega_1^2 + \omega_3^1) = k\phi^1$, and the smooth function k is the seventh order invariant. Thus $u^*\omega_1^2 = \frac{1}{2}k\phi^1 = u^*\omega_3^1$. Taking the exterior derivative of these equations and applying the Maurer-Cartan equations, it follows that $dk = k_{;1}\phi^1 - 2k\phi^2$.

A complete set of local differential invariants of f is given, with respect to the sixth order frame field u along f, by $\{\phi^1, \phi^2, r, k\}$. The structure equations of f are: $d\phi^1 = -\phi^1 \wedge \phi^2$, $d\phi^2 = 0$, $dr = -3r\phi^2$, $r > 0$, and $dk = k_{;1}\phi^1 - 2k\phi^2$. By the third equation r is non-constant. The fine structure of f is:

Type IIbIIIbVaVIbi. r and k are independent, the Frenet frame has order 7;

Type IIbIIIbVaVIbii. k is a function of r, and thus $k = cr^{2/3}$, where $c > 0$ is constant. Theorem 5 of section I.12 applies, and if f is of class C^ω, then $f(S)$ is the union of H-orbits of the points of a profile curve in S, where H is the 1-dimensional analytic subgroup of G whose Lie algebra is $\mathfrak{h} = \{\theta^3 = 0$, $\theta^2 = 0$, $\omega_1^3 = \theta^1$, $\omega_2^3 = 0$, $\omega_2^1 = \theta^1$, $\omega_1^1 = 0$, $\omega_2^2 = 0$, $\omega_1^2 = \frac{1}{2}r_0^{2/3}\theta^1$, $\omega_3^1 = \omega_1^2$, $\omega_3^2 = r_0\theta^1\}$, where $r_0 \in r(S)$ is such that for some $s_0 \in r^{-1}\{r_0\}$, $f(s_0) = 0$ and $u(s_0) = 1$. (Always possible after translating f by an element of G, if necessary).

This completes the Type IIbIIIbVa surfaces.

Type IIbIIIbVb.

We say that a Type IIbIIIb surface $f: S \to M$ is of Type IIbIIIbVb if $m \neq 0$; i.e., $u^*\omega_1^1 \neq 0$ for any fourth order frame field u along f . Then there exists a fourth order frame field u along f such that $t = 0$; i.e., $u^*(\omega_3^2 - m(\omega_1^2 + \omega_3^1)) = r\phi^1$. This condition characterizes the fifth order frame fields along f . The smooth function r on S is a fifth order invariant of f .

The isotropy subgroup of $G_4(m)$ at any point of W_5 is $G_5 = \{1\}$. Thus all higher order frames are the same as the fifth order frames on f ; i.e., $L_j = L_5$ for all $j \geq 5$. The Lie algebra of G_5 is $\mathfrak{g}_5 = (0)$, so $\Theta_5 = \omega_4 = \tau\begin{pmatrix}0 & 0 & 1\\ 1 & 0 & 2m\\ 0 & 0 & 0\end{pmatrix}$, where $\tau = \frac{1}{2}(\omega_1^2 + \omega_3^1)$.

Let u be the fifth order frame field along f , and set $u^*\theta^1 = \phi^1$, $u^*\theta^2 = \phi^2$. Then $u^*\theta^3 = 0$, $u^*\omega_1^3 = \phi^1$, $u^*\omega_2^3 = 0$, $u^*\omega_2^1 = \phi^1$, $u^*\omega_1^1 = -m\phi^1$, $u^*\omega_2^2 = 3m\phi^1 + \phi^2$, $u^*(\omega_1^2 - \omega_3^1) = m\phi^2$, and $u^*(\omega_3^2 - m(\omega_1^2 + \omega_3^1)) = r\phi^1$. We set $u^*(\omega_1^2 + \omega_3^1) = k\phi^1 + s\phi^2$, so the smooth functions k and s on S are sixth order invariants of f . Since $m_{;1} = 0$ and $m_{;2} = -2m$, we have $dm = -2m\phi^2$, i.e., $\phi^2 = d \log|m|^{-1/2}$, and, differentiating, it follows that $s = 5m$.

A complete set of local differential invariants of f is given by $\{\phi^1, \phi^2, m, r, k\}$. The structure equations of f are: $d\phi^1 = -\phi^1 \wedge \phi^2$, $d\phi^2 = 0$, $dm = -2m\phi^2$, $r_{;2} = -3r + \frac{1}{2} mk + 11m^3$, and $k_{;2} = 14m^2 - 2k$, where $dr = r_{;1}\phi^1 + r_{;2}\phi^2$ and $dk = k_{;1}\phi^1 + k_{;2}\phi^2$.

Since $m \neq 0$, it follows from the third structure equation that m is non-constant. If r is a function of m , then $r_{;1} = 0$ and the fourth structure equation implies k is a function of m . The fine structure of this Type is:

<u>Type IIbIIIbVbi</u>. m and r are independent, i.e., $r_{;1} \neq 0$, and the Frenet frame has order 5;

<u>Type IIbIIIbVbii</u>. all invariants are a function of m , and the Frenet frame has order 5. Theorem 5 of section I.12 applies. We omit the details.

This completes the Type IIbIIIb surfaces, and thus completes the analysis of the Type IIb surfaces (the parabolic surfaces).

<u>Type IIc</u> (elliptic surfaces).

We say that a surface $f: S \to M$ is of Type IIc if, at least locally, there exists a first order frame field u along f such that $u^*\omega_1^3 = \phi^1$ and $u^*\omega_2^3 = \phi^2$; i.e., $X = \begin{pmatrix} 1 & 0 \\ 0 & 1 \end{pmatrix}$. This condition characterizes the second order frame fields along f . There are no second order invariants.

The isotropy subgroup of G_1 at $\begin{pmatrix} 1 & 0 \\ 0 & 1 \end{pmatrix}$ is G_2

$$= \left\{ \begin{pmatrix} A & \xi \\ 0\ 0 & a \end{pmatrix} \in G_1 : a^{-1}\, {}^tAIA = I \right\} = \left\{ \begin{pmatrix} A & \xi \\ 0\ 0 & 1 \end{pmatrix} \in G_1 : \xi \in \mathbb{R}^2 , \ A \in SO(2) \right\} ,$$

which is connected. Its Lie algebra is $\mathfrak{g}_2 = \left\{ \begin{pmatrix} 0 & -x & y \\ x & 0 & z \\ 0 & 0 & 0 \end{pmatrix} : x, y, z \in \mathbb{R} \right\}$,

and for a subspace of $\mathfrak{g}_1$ complementary to $\mathfrak{g}_2$ we take

$$\mathfrak{m}_2 = \left\{ \begin{pmatrix} x & s & 0 \\ s & y & 0 \\ 0 & 0 & t \end{pmatrix} : x, y, s, t \in \mathbb{R} , \quad x + y + t = 0 \right\} .$$ The decomposition $\mathfrak{g}_1 = \mathfrak{g}_2 + \mathfrak{m}_2$ decomposes ω_1 into $\omega_1 = \omega_2 + \Theta_2$, where the $\mathfrak{m}_2$-component is $\Theta_2 = \begin{pmatrix} \omega_1^1 & \nu & 0 \\ \nu & \omega_2^2 & 0 \\ 0 & 0 & \omega_3^3 \end{pmatrix}$, $\nu = \frac{1}{2}(\omega_2^1 + \omega_1^2)$, and ω_2 is the $\mathfrak{g}_2$-component.

Let u be a second order frame field along f , let $u^*\theta^1 = \phi^1$, $u^*\theta^2 = \phi^2$, and set: $u^*\omega_1^1 = x\phi^1 + s\phi^2$, $u^*\omega_2^2 = y\phi^1 + t\phi^2$, and $u^*(\omega_2^1 + \omega_1^2) = p\phi^1 + q\phi^2$, where x, y, s, t, p, q are smooth functions on S . Taking the exterior derivative of $u^*\omega_1^3 = \phi^1$ and $u^*\omega_2^3 = \phi^2$, and applying the Maurer-Cartan equations of G , we get: $p = 3s + t$ and $q = x + 3y$. If we set $X = \begin{pmatrix} x & \frac{1}{2}(3s+t) \\ \frac{1}{2}(3s+t) & y \end{pmatrix}$ and $Y = \begin{pmatrix} s & \frac{1}{2}(x+3y) \\ \frac{1}{2}(x+3y) & t \end{pmatrix}$, then $u^*\begin{pmatrix} \omega_1^1 & \nu \\ \nu & \omega_2^2 \end{pmatrix} = (X, Y)\begin{pmatrix} \phi^1 \\ \phi^2 \end{pmatrix}$, where $\nu = \frac{1}{2}(\omega_2^1 + \omega_1^2)$ as above.

Any other second order frame field along f is given by $\tilde{u} = uK$, where $K = \begin{pmatrix} A & \xi \\ 0\ 0 & 1 \end{pmatrix}$, $A\colon S \to SO(2)$, $\xi\colon S \to \mathbb{R}^2$ smooth maps. Let $\tilde{\phi}^1, \tilde{\phi}^2, \tilde{x}, \tilde{y}, \tilde{s}, \tilde{t}, \tilde{X}, \tilde{Y}$ be the analogs of $\phi^1, \phi^2, x, y, s, t, X, Y$, respectively, for $\tilde{u}$. Now $\tilde{u}^*(\Theta_0 + \Theta_1 + \Theta_2) =$ the $(\mathfrak{m}_0 + \mathfrak{m}_1 + \mathfrak{m}_2)$-component of $\operatorname{ad}(K^{-1})u^*(\Theta_0 + \Theta_1 + \Theta_2)$ implies that $\begin{pmatrix} \tilde{\phi}^1 \\ \tilde{\phi}^2 \end{pmatrix} = A^{-1}\begin{pmatrix} \phi^1 \\ \phi^2 \end{pmatrix}$ and

$$\text{and}\quad u^*\begin{pmatrix}\omega_1^1 & \nu \\ \nu & \omega_2^2\end{pmatrix} = A^{-1}u^*\begin{pmatrix}\omega_1^1 & \nu \\ \nu & \omega_2^1\end{pmatrix}A + \frac{1}{2}\left\{(-A^{-1}\xi)(\phi^1, \phi^2) + \begin{pmatrix}\phi^1 \\ \phi^2\end{pmatrix}^t(-A^{-1}\xi)\right\}.$$

Combining these we get:

$$(*)\quad (\tilde{X}, \tilde{Y})$$

$$= (A^{-1}XA,\ A^{-1}YA)A$$

$$+ \frac{1}{2}((-A^{-1}\xi)^t(A^{-1}\varepsilon_1) + (A^{-1}\varepsilon_1)^t(-A^{-1}\xi),\ (-A^{-1}\xi)^t(A^{-1}\varepsilon_2) + (A^{-1}\varepsilon_2)^t(-A^{-1}\xi))A\ ,$$

where $\varepsilon_1, \varepsilon_2$ is the standard basis of $\mathbb{R}^2$.

Let $V = \{(X, Y) \in \mathcal{S}_2 \colon Y_{12} = \frac{1}{2}(X_{11} + 3X_{22})\ ,\ X_{12} = \frac{1}{2}(3Y_{11} + Y_{22})\}$ $= \left\{\left(\begin{pmatrix} x & \frac{1}{2}(3s+t) \\ \frac{1}{2}(3s+t) & y\end{pmatrix}, \begin{pmatrix} s & \frac{1}{2}(x+3y) \\ \frac{1}{s}(x+3y) & t\end{pmatrix}\right) : x, y, s, t \in \mathbb{R}\right\}$. Define an affine action ρ of G_2 on V by $\rho(A, \xi)(X, Y)$ $= (AXA^{-1}, AYA^{-1})A^{-1} + \frac{1}{2}(\xi^t(A\varepsilon_1) + A\varepsilon_1{}^t\xi,\ \xi^t(A\varepsilon_2) + A\varepsilon_2{}^t\xi)A^{-1}$, where $A \in SO(2)$, $\xi \in \mathbb{R}^2$. (This comes from $(*)$ since $K^{-1} = \begin{pmatrix} A^{-1} & -A^{-1}\xi \\ 0\ 0 & 1\end{pmatrix}$.)

The proofs of the following Lemmas are elementary and straightforward, and thus are only sketched or omitted.

<u>Lemma 3</u>. If $(X, Y) \in V$, $A \in SO(2)$, $\xi \in \mathbb{R}^2$ and (X^1, Y^1) $= \rho(A, \xi)(X, Y)$, then $(\mathrm{Tr}\ X^1, \mathrm{Tr}\ Y^1) = (\mathrm{Tr}\ X, \mathrm{Tr}\ Y)A^{-1} + {}^t\xi$, where $\mathrm{Tr} = \mathrm{Trace}$.

Corollary. Every G_2 orbit in V meets the subspace
$W = \{(X, Y) \in V: \text{Tr } X = 0 = \text{Tr } Y\} = \{(\begin{pmatrix} x & y \\ y & -x \end{pmatrix}, \begin{pmatrix} y & -x \\ -x & -y \end{pmatrix}): x, y \in \mathbb{R}\}$.

Proof: Take $(X, Y) \in V$, let $A = I$ and $^{*}\xi = -(\text{Tr } X, \text{Tr } Y)$.
Then $\rho(A, \xi)(X, Y) \in W$.

Lemma 4. Consider the vector space isomorphism $W \cong \mathbb{R}^2$ given by
$W \ni (\begin{pmatrix} x & y \\ y & -x \end{pmatrix}, \begin{pmatrix} y & -x \\ -x & -y \end{pmatrix}) \leftrightarrow \begin{pmatrix} x \\ y \end{pmatrix} \in \mathbb{R}^2$. Then for any $A \in SO(2)$,
$\rho(A, 0)W \subseteq W$ and

$$\rho(A, 0)(\begin{pmatrix} x & y \\ y & -x \end{pmatrix}, \begin{pmatrix} y & -x \\ -x & -y \end{pmatrix}) \leftrightarrow A^3\begin{pmatrix} x \\ y \end{pmatrix} .$$

Proof: Compute directly from (*) with $\xi = 0$.

Lemma 5. A cross section for the action of G_2 on V is
$W_2 = \{r(\begin{pmatrix} 1 & 0 \\ 0 & -1 \end{pmatrix}, \begin{pmatrix} 0 & -1 \\ -1 & 0 \end{pmatrix}): r \geq 0\}$.

Proof: First check that the intersection of W with a G_2 orbit of V is precisely one $SO(2)$ orbit of W , i.e., if $(X, Y) \in V$, then $W \cap \rho(G_2)(X, Y) = \rho(SO(2), 0)(X', Y')$, for any $(X', Y') \in W \cap \rho(G_2)(X, Y)$
Now it follows from Lemma 4 that any $SO(2)$ orbit in W meets W_2 in exactly one point.

Remark. The G_2 orbit of the origin $(0, 0)$ is the linear subspace
$N = \{(\begin{pmatrix} 2t & s \\ s & 0 \end{pmatrix}, \begin{pmatrix} 0 & t \\ t & 2s \end{pmatrix}): t, s \in \mathbb{R}\}$.

<u>Lemma 6</u>. Suppose $(X, Y): S \to V$ is a smooth map which either (i) never meets N, or (ii) lies in N. Then there exists, locally, a smooth map $K = \begin{pmatrix} A & \xi \\ 0\ 0 & 1 \end{pmatrix}: S \to G_2$ such that $\rho(A, \xi)(X, Y)$ lies in $W_2 - \{(0, 0)\}$ in case (i), and is identically equal to $(0, 0)$ in case (ii).

<u>Proof</u>: There exists a smooth map $\xi: S \to \mathbb{R}^2$, for example ${}^t\xi = -(\operatorname{Tr} X, \operatorname{Tr} Y)$, such that $\rho(I, \xi)(X, Y)$ lies in W. Thus it suffices to prove the lemma when (X, Y) lies in W. But this is just the case of a compact Lie group action with slice W_2.

Thus we have two cases, depending on whether the map $(X, Y): S \to B$ lies in N, (Type IIcIIIa), or lies in the complement of N, (Type IIcIIIb).

<u>Type IIcIIIa</u>. (Pick invariant equal to zero.)

There are the surfaces with Pick invariant zero as discussed by Spivak in [S, 1975], p. 171. We say that a Type IIc surface $f: S \to M$ is of Type IIcIIIa if the above map $(X, Y): S \to V$ takes all values in N, for any second order frame field u along f. Then by Lemma 6, there exists (at least locally) a second order frame field u such that $(X, Y) = (0, 0)$, where $X = \begin{pmatrix} x & \frac{1}{2}(3s+t) \\ \frac{1}{2}(3s+t) & y \end{pmatrix}$,

$Y = \begin{pmatrix} s & \frac{1}{2}(x+3y) \\ \frac{1}{2}(x+3y) & t \end{pmatrix}$, $u^*\omega_1^1 = x\phi^1 + s\phi^2$, $u^*\omega_2^2 = y\phi^1 + t\phi^2$ and

$u^*(\omega_2^1 + \omega_1^2) = (3s + t)\phi^1 + (x + 3y)\phi^2$. Hence $x = y = s = t = 0$; i.e., $u^*\omega_1^1 = u^*\omega_2^2 = u^*(\omega_2^1 + \omega_1^2) = 0$. This condition characterizes the third order frame fields along f . There are no third order invariants.

The isotropy subgroup of G_2 at $(0, 0) \in W_2$ is

$$G_3 = \left\{ \begin{pmatrix} A & & 0 \\ & & 0 \\ 0 & 0 & 1 \end{pmatrix} : A \in SO(2) \right\}.$$ Its Lie algebra is

$$\mathcal{G}_3 = \begin{pmatrix} 0 & -z & 0 \\ z & 0 & 0 \\ 0 & 0 & 0 \end{pmatrix} : z \in \mathbb{R}$$, and for a subspace of $\mathcal{G}_2$ complementary to $\mathcal{G}_3$ we take $$\mathcal{M}_3 = \left\{ \begin{pmatrix} 0 & & r \\ & & t \\ 0 & 0 & 0 \end{pmatrix} : r, t \in \mathbb{R} \right\}.$$ Then ω_2 , the $\mathcal{G}_2$-component of Ω , decomposes into $\omega_2 = \omega_3 + \Theta_3$, where the $\mathcal{M}_3$-component is $$\Theta_3 = \begin{pmatrix} 0 & 0 & \omega_3^1 \\ 0 & 0 & \omega_3^2 \\ 0 & 0 & 0 \end{pmatrix},$$ and ω_3 is the $\mathcal{G}_3$-component.

Let u be a third order frame field along f , let $\phi^1 = u^*\theta^1$ and $\phi^2 = u^*\theta^2$, and then $u^*\omega_1^1 = 0$, $u^*\omega_2^2 = 0$, and $u^*(\omega_2^1 + \omega_1^2) = 0$ Taking the exterior derivative of these last three equations, and applying the Maurer-Cartan equations of G , it follows that $u^*\omega_3^1 = k\phi^1$ and $u^*\omega_3^2 = k\phi^2$ for some smooth function k on S .

Any other third order frame field along f is given by $\tilde{u} = uK$, where $$K = \begin{pmatrix} A & & 0 \\ & & 0 \\ 0 & 0 & 1 \end{pmatrix},$$ $A: S \to SO(2)$ is a smooth map. Let $\tilde{\phi}^1, \tilde{\phi}^2, \tilde{k}$

be the analogues of ϕ^1, ϕ^2, k for $\tilde{u}$. Then $\tilde{u}^*(\Theta_0 + \ldots + \Theta_3)$ $= (\mathfrak{m}_0 + \ldots + \mathfrak{m}_3)$-component of $\mathrm{ad}(K^{-1})u^*(\Theta_0 + \ldots + \Theta_3)$ implies that $(\tilde{\phi}^1, \tilde{\phi}^2) = (\phi^1, \phi^2)A$ and $\tilde{k} = k$. Taking the exterior derivative of $u^*\omega_3^1 = k\phi^1$ and $u^*\omega_3^2 = k\phi^2$, and applying the Maurer-Carton equations of G, it follows that $k_{;1} = 0 = k_{;-}$, that is, k is constant. The constant k is a fourth order invariant.

All higher order frames are the same as the third order frames; i.e., $L_j = L_3$, for all $j \geqslant 3$.

Since the only invariant is constant on S, we know that $f(S)$ must be part of a homogeneous surface. For each $k \in \mathbb{R}$, consider the 3-dimensional, left-invariant, involutive distribution $\mathfrak{h}_k$ on G defined by the equations $\{\theta^3 = 0,\ \omega_1^3 = \theta^1,\ \omega_2^3 = \theta^2,\ \omega_1^1 = 0,\ \omega_2^2 = 0,\ \omega_2^1 + \omega_1^2 = 0,\ \omega_3^1 = k\theta^1,\ \omega_3^2 = k\theta^2\}$. Then

$$\mathfrak{h}_k = \left\{\left(\begin{pmatrix} X & k\xi \\ {}^t\xi & 0 \end{pmatrix}, \begin{pmatrix} \xi \\ 0 \end{pmatrix}\right) : X \in \mathfrak{o}(2),\ \xi \in \mathbb{R}^2\right\}.$$

The analytic subgroup of G whose Lie algebra is $\mathfrak{h}_k$ is H_k, where: for $k > 0$, $H_k = \{(A, (I_k^{-1} + k^{-1}A)\varepsilon_3) : {}^tAI_kA = I_k,\ \det A = 1,\ {}^t\varepsilon_3A\varepsilon_3 \geqslant 1\}$; for $k < 0$, $H_k = \{(A, (I_k^{-1} + k^{-1}A)\varepsilon_3) : {}^tAI_kA = I_k,\ \det A = 1\}$;

$$\text{for } k = 0,\quad H_0 = \left\{\left(\begin{pmatrix} a & 0 \\ & 0 \\ {}^t\xi & 1 \end{pmatrix}, \begin{pmatrix} a\xi \\ \\ \frac{1}{2}\,{}^t\xi\xi \end{pmatrix}\right) :\ a \in SO(2),\ \xi \in \mathbb{R}^2\right\}.$$

Here

$$I_k = \begin{pmatrix} 1 & 0 & 0 \\ 0 & 1 & 0 \\ 0 & 0 & -k \end{pmatrix},\quad \varepsilon_3 = \begin{pmatrix} 0 \\ 0 \\ 1 \end{pmatrix}.$$

Theorem. Let $f\colon S \to \mathbb{R}^3$ be a Type IIcIIIa surface with (constant) fourth order invariant k. Then $f(S)$ is G-congruent to an open submanifold of the homogeneous Type IIcIIIa surface h_k, where h_k is: the ellipsoid of revolution $-kx^2 - ky^2 + (kz + 1)^2 = 1$ if $k < 0$; the upper sheet of the two-sheeted hyperboloid of revolution $-kx^2 - ky^2 + (kz + 1)^2 = 1$ if $k > 0$; the parabaloid of revolution $2z = x^2 + y^2$ if $k = 0$. Each of these surfaces is, in fact, a Type IIcIIIa surface the value of whose invariant is the number k associated with that equation.

Proof: Let $u\colon S \to G$ be a third order frame field along f. Using the notation just preceding this Theorem, it follows from the characterizing properties of such a frame field along f that $u(S)$ is an integral submanifold of the distribution $\mathcal{G}_k$. Thus a G-translate of $u(S)$ is contained in H_k. Translating $f(S)$ by that amount, we may as well assume $u(S) \subset H_k$. Hence, after a G-translate if necessary, $f(S) = \pi u(S) \subset \pi H_k$, and it is easily verified that $\pi H_k = h_k$.

Type IIcIIIb.

We say that a Type IIc surface $f\colon S \to M$ is of Type IIcIIIb if the above map $(X, Y)\colon S \to V$ takes values in the complement of N in V, for any second order frame field u along f. Then by Lemma 6, there exists a second order frame field u such that $(X, Y) \in W_2 - \{(0, 0)\}$; i.e., $(X, Y) = \left(\begin{pmatrix} x & 0 \\ 0 & -x \end{pmatrix}, \begin{pmatrix} 0 & -x \\ -x & 0 \end{pmatrix}\right)$, $x > 0$, so that $u^*\omega_1^1 = x\phi^1$, $u^*\omega_2^2 = -x\phi^1$, and $u^*(\omega_2^1 + \omega_1^2) = -2x\phi^2$, for

$x > 0$. This condition characterizes the third order frame fields along f . The smooth positive function x on S is the third order invariant of f . It is called the Pick invariant by Spivak in [S, 1975], p. 171.

The isotropy subgroup of G_2 at any point in $W_2 \setminus \{(0, 0)\}$ is

$$G_3 = \left\{ \begin{pmatrix} A & \begin{matrix} 0 \\ 0 \end{matrix} \\ 0\ 0 & 1 \end{pmatrix} : A \in SO(2),\ A^3 = I \right\} ,$$ which has three elements, corresponding to three orientation classes of third order frames. Since

$$\mathfrak{g}_3 = (0) , \quad \text{we have} \quad \Theta_3 = \omega_2 = \begin{pmatrix} 0 & -\sigma & \omega_3^1 \\ \sigma & 0 & \omega_3^2 \\ 0 & 0 & 0 \end{pmatrix} , \quad \sigma = \frac{1}{2}(\omega_1^2 - \omega_2^1) .$$

Let u be a third order frame field along f and set $\phi^1 = u^*\theta^1$, $\phi^2 = u^*\theta^2$. Then $u^*\theta^3 = 0$, $u^*\omega_1^3 = \phi^1$, $u^*\omega_2^3 = \phi^2$, $u^*\omega_1^1 = x\phi^1$, $u^*\omega_2^2 = -x\phi^1$, $u^*(\omega_2^1 + \omega_k^2) = -2x\phi^2$. Taking the exterior derivative of the last three equations, and applying the Maurer-Cartan equations of G , it follows that $\phi^1 \wedge u^*\omega_3^1 = u^*\omega_3^2 \wedge \phi^2$. Thus $u^*\omega_3^1 = z\phi^1 + c\phi^2$, $u^*\omega_3^2 = c\phi^1 + k\phi^2$ and, if we set $u^*(\omega_2^1 - \omega_1^2) = a\phi^1 + b\phi^2$, then a, b, c, k, z are smooth functions on S satisfying $2x_{;1} = 3xb + z - k$ and $2x_{;2} = -3xa - 2c$. These five new functions are fourth order invariants of f .

Again taking the exterior derivative, this time of the expressions for $u^*\omega_3^1$, $u^*\omega_3^2$ and $u^*(\omega_2^1 - \omega_1^2)$, and applying the Maurer-Cartan equations, we obtian the remaining relations on the invariants, which we collect together to get the structure equations of f :

$$d\phi^1 = -\frac{1}{2} a\phi^1 \wedge \phi^2 , \qquad d\phi^2 = -\frac{1}{2} b\phi^1 \wedge \phi^2 ,$$

$$2x_{;1} = 3xb + z - k , \qquad 2x_{;2} = -3xa - 2c ,$$

$$b_{;1} - a_{;2} = 4x^2 + \frac{1}{2}(a^2 + b^2) - z - k ,$$

$$c_{;1} - z_{;2} = c(b - 2x) + \frac{1}{2} a (z - k) ,$$

and

$$k_{;1} - c_{;2} = ac + \frac{1}{2}(k - z)(b + 2x) .$$

Any other third order frame field along f is given by $\tilde{u} = uK$, where $K = \begin{pmatrix} A & \begin{matrix} 0 \\ 0 \end{matrix} \\ 0\ 0 & 1 \end{pmatrix}$, $A \in SO(2)$ with $A^3 = I$. Let $\tilde{\phi}^1, \tilde{\phi}^2, \tilde{a}, \tilde{b}, \tilde{c}, \tilde{k}, \tilde{z}$ be the analogs of these quantities for $\tilde{u}$. Then $\tilde{u}^*\Omega = \text{ad}(K^{-1})u^*\Omega$ implies that $(\tilde{\phi}^1, \tilde{\phi}^2) = (\phi^1, \phi^2)A$, $(\tilde{a}, \tilde{b}) = (a, b)^tA$, and $\begin{pmatrix} \tilde{z} & \tilde{c} \\ \tilde{c} & \tilde{k} \end{pmatrix} = {}^tA\begin{pmatrix} z & c \\ c & k \end{pmatrix}A$. There are two orbit types: either $a = b = x_{;1} = x_{;2} = 0$, in which case $z = k = 2x^2$ is a positive constant and $c = 0$ by the third, fourth and fifth structure equations, (Type IIcIIIbIVa); or at least one of $a, b, x_{;1}, x_{;2}$ is non-zero, (Type IIcIIIbIVb).

<u>Type IIcIIIbIVa.</u>

We say a Type IIcIIIb surface $f: S \to M$ is of Type IIcIIIbIVa if $a = b = x_{;1} = x_{;2} = 0$; i.e., $x > 0$ is constant and $u^*(\omega_2^1 - \omega_1^2) = 0$ for any third order frame field u. As pointed out above, it follows that $z = k = 2x^2$ and $c = 0$. Hence fourth order frames are the same as third order frames, and in fact $L_j = L_3$ for all $j \geqslant 3$, because there are no non-constant invariants.

The Corollary of Theorem 5 of section I.12 applies. Any Type IIcIIIbIVa surface is G-congruent to an open submanifold of the homogeneous Type IIcIIIbIVa surface $\pi H(x)$, where, for each $x > 0$, $H(x)$ is the analytic subgroup of the 2-dimensional, abelian Lie algebra $\mathfrak{h}(x)$ of $\mathfrak{g}$ given by $\mathfrak{h}(x) = \{\theta^3 = 0 \, , \; \omega_1^3 = \theta^1 \, , \; \omega_2^3 = \theta^2 \, ,$ $\omega_1^1 = x\theta^1 \, , \; \omega_2^2 = -x\theta^1 \, , \; \omega_2^1 = -x\theta^1 \, , \; \omega_1^2 = -x\theta^1 \, , \; \omega_3^1 = 2x^2\theta^1 \, ,$

$$\omega_3^2 = 2x^2\theta^2\} = \left\{\left(\begin{pmatrix} xr & -xt & 2x^2r \\ -xt & -xr & 2x^2t \\ r & t & 0 \end{pmatrix}, \begin{pmatrix} r \\ t \\ 0 \end{pmatrix}\right) : r, t \in \mathbf{R}\right\}.$$

The positive constant x is the third order invariant.

Type IIcIIIbIVb.

We say a Type IIcIIIb surface $f\colon S \to M$ is of Type IIcIIIbIVb if at least one of $a, b, x_{;1}, x_{;2}$ is non-zero; i.e., x is non-constant or $u^*(\omega_2^1 - \omega_1^2) \neq 0$ for any third order frame field u . Then the fourth order frames of f are a specific choice of one of the three orientation classes of L_3 . We shall not bother to make the choice explicit. The smooth functions $x_{;1}, x_{;2}, a, b, c, k, z$ on S are the fourth order invariants of f .

The isotropy subgroup of G_3 is $G_4 = \{1\}$, and $L_j = L_4$, for all $j \geqslant 3$. We shall omit an analysis of the fine structure of this type surface, except to point out that there are two familes of homogeneous surfaces of this type. Namely, those for which the constant invariants are:

(i) $a = c = k = 0$, $b = -2x$, $z = 6x^2$, $x > 0$; and

(ii) $a = \varepsilon\sqrt{3}x$, $b = x$, $c = \varepsilon \frac{\sqrt{3}}{2} x^2$, $k = \frac{9}{2} x^2$, $z = \frac{3}{2} x^2$, $x > 0$, $\varepsilon = \pm 1$.

This completes the analysis of Type IIc surfaces.

REFERENCES

[B, 1972] Bredon, G.E. Introduction to Compact Transformation Groups, Academic Press, New York, 1972.

[C, 1935] Cartan, E. La Méthode du Repère Mobile, la Théorie des Groups Continus, et les Espaces Généralisés, Hermann, Paris, 1935.

[C, 1937] ________. Théorie des Groupes Finis et Continus et la Géométrie Différentielle traitées par la Méthode du Repère Mobile, Gauthier-Villars, Paris, 1937.

[G, 1963] Guggenheimer, H.W. Differential Geometry, McGraw-Hill, New York, 1963.

[G, 1974] Griffiths, P. On Cartan's method of Lie groups and moving frames as applied to uniqueness and existence questions in differential geometry, Duke Math. Journ. 41, (1974), 775-814.

[K, 1926] Klein, F. Vorlesungen über Höhere Geometrie, dritte Auflage, Springer, Berlin, 1926.

[S, 1975] Spivak, M. A Comprehensive Introduction to Differential Geometry, Vol. III, Publish or Perish, Inc., Boston, 1975.

[W, 1938] Weyl, H. Cartan on groups and differential geometry, Bulletin, Amer. Math. Soc. 44(1938), 598-601.

INDEX

Adapted frames 6

Adjoint representation 7

Bundle map 6, 13, 18

$\mathbb{C}G_{4,2}$ 81

$\mathbb{C}P^2$ 67

Calabi's Theorem 79

Congruence, definition 5
  theorem in C^ω case 23
    in C^{q+1} case 31
    in Lie groups 30

Contact, definition 1
  between graphs 2
  G-contact 5, 10

Covariant derivative 27

Curves in $\tilde{G}_{4,2} = \frac{SO(4)}{SO(2)\times SO(2)}$ 58
  Type: Ia 60; IaIIa ("helix") 61; IaIIb ("helix") 62; Ib 62; Ic 63; IcIIa ("geodesic") 64; IcIIb ("helix") 64

$\mathbb{E}(3)$ (proper rigid motions) 45

Existence of submanifolds 30

Frame field 6
  zeroth order 7; first order 9;
  second order 14; third order 20

Frenet frame 22

$\tilde{G}_{4,2}$ 5

Gauss curvature 78

Grassmann bundle 5; of $\mathbb{R}^n$ 3
  manifold 23, 58

$h_0: G \to L(M)$ 6;
  $h_1: G \times W_1 \to L(M^{(1)})$ 13;
  $h_2: G \times W_2 \to L(M^{(2)})$ 18

Holomorphic curves in $\mathbb{C}G_{4,2}$ 81
  Type: Ia 83; IaIIa (special ruled surface) 85; IaIIb 86; IaIIbIIIa (pencil) 88; IaIIbIIIb 91; IaIIc 93; IaIIcIIIa (homogeneous) 95; IaIIcIIIb 97; Ib (special ruled surface) 98; Ic 101; IcIIa (special ruled surfaces) 103; IcIIb (homogeneous) 106; IcIIc 110; IcIId 113.

Holomorphic curves in $\mathbb{C}P^2$ 66
  Type: IIa (hyperplane) 71; IIb 71; IIbIIIa (Veronese variety) 73; IIbIIIb 74; IIbIIIbi (generic) 76; IIbIIIbii (W-curves) 76

Homogeneous submanifolds 41

Invariants: first order 10; second order 14; third order 20; in terms of Maurer-Cartan form 25; constant 42

Isotropy subgroup 6

L_0 7; L_1 9; L_2 14; L_3 20

$\lambda_0: L_0 \to G_{m_0,p}$ 5, 8;

$\lambda_1: L_1 \to G_{m_1+\mu_1,p}$ 14;

$\lambda_2: L_2 \to G_{m_2+\mu_2,p}$ 19

Linear isotropy representation 6

Local cross section 8,14

Maurer-Cartan form 23

equations 23

Orientation of frames 7,10, 15

Pick invariant 143, 147

Projections $\psi_1: G_{m_1+\mu_1;p} \to G_{m_0,p}$ 14;

$\psi_2: G_{m_2+\mu_2;p} \to G_{m_1+\mu_1;p}$ 19

$\mathbb{R}^{n\times p}$, $\mathbb{R}^{n\times p*}$ 23

$\rho_0: G_0 \to GL(m_0, \mathbb{R})$ 6;

$\rho_1: G_1 \to GL(m_1 + \mu_1; \mathbb{R})$ 13

$\rho_2: G_2 \to GL(m_2 + \mu_2; \mathbb{R})$ 19

Relations among invariants 26

SO(4) ; SU(3) ; SU(4) 81

Special affine group $SL(3;\mathbb{R})\cdot\mathbb{R}^3$ 116

Special affine surface theory 115

Type: IIa (planar) 120;
IIb (parabolic) 120;
IIbIIIa 122; IIbIIIaIVa (cylinder $z = \frac{1}{2}x^2$) 124;
IIbIIIaIVb 124; IIbIIIaIVbi (homogeneous cylinder) 126;
IIbIIIaIVbii (nonhomogeneous cylinders) 127;
IIbIIIb 129;
IIbIIIbVa 133; IIbIIIbVaVIa (homogeneous cone) 134;
IIbIIIbVaVIb 136;
IIbIIIbVb 138; IIc (elliptic) 139; IIcIIIa (homogeneous quadric) 143;
IIcIIIb 146; IIcIIIbIVa (group manifold) 148,
IIcIIIbIVb 149

Special ruled surfaces in $\mathbb{C}P^3$ 85 f, 98-100, 103-106

Structure equations of f 40

Surfaces in Euclidean space 44

Type: IIa (plane) 49; IIb (spheres) 50; IIc (minimal surfaces) 51; IId (generic) 52; IIdIIIa (right circular cylinder) 55; IIcIIIiii and IIdIIIbiii (cylinders, surfaces of revolution, helicoids) 57

Taylor polynomial 3, 4

Type of a cross section 9, 14, 20;
constant type 12, 21

W-curves of Klein and Lie 76

Weingarten surfaces 57